Evacuation Modeling Trends

Arturo Cuesta • Orlando Abreu • Daniel Alvear
Editors

Evacuation Modeling Trends

Springer

Editors
Arturo Cuesta
GIDAI Group
University of Cantabria
Santander, Cantabria, Spain

Orlando Abreu
GIDAI Group
University of Cantabria
Santander, Cantabria, Spain

Daniel Alvear
GIDAI Group
University of Cantabria
Santander, Cantabria, Spain

ISBN 978-3-319-20707-0 ISBN 978-3-319-20708-7 (eBook)
DOI 10.1007/978-3-319-20708-7

Library of Congress Control Number: 2015951779

Springer Cham Heidelberg New York Dordrecht London

Printed on acid-free paper

Springer International Publishing AG Switzerland is part of Springer Science+Business Media (www.springer.com)

Preface

Nobody doubts how important the evacuation process for life safety can be. Evacuation modelling essentially aims to understand this complex phenomenon. This requires identifying and selecting the relevant aspects of this process in the real world and then using a model (that usually include a set of conceptual, mathematical and, frequently, computational models) to represent them. Despite being a simplified reflection of reality, evacuation modelling is very useful because it helps us to simulate, visualize, manipulate and gain additional information about the process being represented and, therefore, improve life safety. Evacuation modelling allows the main features of this process to be quantified and thus contributes to their optimization. This book, *Evacuation Modeling Trends*, is intended for those involved in safety, from emergency and intervention personnel to students, engineers and researchers. The chapter authors are international experts in their own fields. The book covers basic concepts and definitions, theory on human behavior and human performance data, available tools and simulation approaches, model development, application and validation methods and suggests some challenges for the future in the subject.

Chapter 1 presents basic concepts and definitions of evacuation modelling, describes the different solutions to represent the evacuees and the geometry and discusses the current evacuation modelling methods.

Chapter 2 discusses the nature and importance of the different parameters involved in ASET and RSET and the interactions between them. Consideration is given to the data and research efforts needed to develop and incorporate functions for the different parameters into comprehensive escape and evacuation simulations, with a number of examples illustrating different aspects and approaches.

Chapter 3 addresses the information available regarding how people typically behave during an evacuation, the tools used to represent this performance and how these tools might be properly implemented. The focus is on individual evacuee performance—what we know about it and how it might be represented.

Chapter 4 discusses the verification and validation (V&V) methods of evacuation modelling, revises current literature on V&V approaches and presents a set of recommended verification tests and examples of validation tests.

Chapter 5 discusses some of the future challenges in evacuation modelling: new scenarios and factors for future model developments, addresses the problem of using deterministic and/or stochastic approaches in evacuation modelling and proposes and discusses the use of evacuation models for supporting timely decisions in real time.

Santander, Spain

Arturo Cuesta
Orlando Abreu
Daniel Alvear

Contents

1 **Basic Concepts and Modelling Methods** ... 1
Enrico Ronchi and Daniel Nilsson

2 **Developments in Tenability and Escape Time Assessment for Evacuation Modelling Simulations** ... 25
David A. Purser

3 **Representation of Human Behaviour in Fire** ... 55
Steven M.V. Gwynne and Erica D. Kuligowski

4 **Verification and Validation Methods** ... 81
Weiguo Song and Libi Fu

5 **Future Challenges in Evacuation Modelling** ... 103
Arturo Cuesta, Orlando Abreu, and Daniel Alvear

Glossary ... 131

Contents

Chapter 1
Basic Concepts and Modelling Methods

Enrico Ronchi and Daniel Nilsson

1.1 Introduction

Try to imagine a fire emergency evacuation. What is the first image that you have in your mind? Here there are some guesses. Probably a parent running inside a house trying to rescue her children, but not managing since she is too scared? An old man stuck crying in a corner of a room with flames and smoke all around him? Or a group of people crushing aggressively against each other in order to get out of a building through a very small exit door? Almost surely you have thought about people shouting and screaming for help, without doing any constructive actions. Well, these scenarios represent a significant minority of what a fire evacuation generally is. When we think about a fire emergency, we often think about people losing their rationality, rushing irrationally in search of an exit or searching desperately for help. Nevertheless, researchers generally do not agree with this interpretation. This misconception comes from the public opinion, in which a fire emergency is often linked to the word "panic". Human behaviour in fire emergencies used in media accounts and survivors' statement is generally associated with panic behaviour. In contrast with this stereotype, the concept that a fire might cause panic has been abandoned by the scientific community [1]. The definition of panic itself has been largely questioned in several research studies [2, 3]. Feelings such as anxiety and stress may occur during a fire emergency, but they do not generally lead to irrational or anti-social behaviours [1]. Evacuees generally tend to behave rationally, with non-rational and anti-social behaviours occurring in very rare occasions, mostly in very extreme scenarios in which the probability of surviving perceived is extremely low [2]. In fact, most people do not develop shock reactions, and tend to

E. Ronchi (✉) • D. Nilsson
Department of Fire Safety Engineering, Lund University,
118, Lund 22100, Sweden
e-mail: enrico.ronchi@brand.lth.se; daniel.nilsson@brand.lth.se

A. Cuesta et al. (eds.), *Evacuation Modeling Trends*,
DOI 10.1007/978-3-319-20708-7_1

act in accordance to what they think is in their best interest, given the limited understanding they have of the situation.

The de-mystification of the misconception about panic is needed in order to talk about evacuation modelling. The reason is very simple: in order to discuss models of evacuation behaviour, the starting point is that evacuation can be considered as a rational process. This idea permits the study of evacuation with theoretical frameworks.

Another common myth is that the "human factor" as such introduces variables that are almost impossible to predict with any degree of certainty. But is this completely true? The statement that we will never be able to predict what we would do during a fire emergency is both true and false. In fact, it is true that it is today almost impossible to exactly predict what a single individual would do in a fire emergency in his/her house or office. Nevertheless, people tend to show certain trends in their behaviours in relation to their internal characteristics (psychological factors, previous experience, etc.) and external factors (environmental conditions, etc.) [3, 4]. For this reason, from a scientific perspective, it is incorrect to assume that human behaviour is not completely predictable; it is simply a matter of assessing the level of uncertainty associated with the predictions that we can make, given the current theoretical knowledge on human behaviour in fire.

Evacuation modelling is therefore a tool to simulate human behaviour which starts from two assumptions, namely:

1) human behaviour during evacuation is rational
2) human behaviour during evacuation can be predicted

1.1.1 Basic Concepts and Definitions

The present section has the objective to help the reader understand the basic concepts associated with evacuation modelling in fire emergencies. It includes a brief description of the context in which evacuation modelling has been developed and discusses the evolution of the methods adopted for modelling human behaviour in fire. Fire evacuation modelling is a field of science which relates to the simulation of human behaviour during fire emergencies. The need for evacuation modelling tools derives from a paradigm shift in the fire safety design of buildings, i.e., the introduction of performance-based design. The performance-based design approach is based on the concept that any fire safety design can be used in a building as long as an adequate level of safety is provided. This is in contrast with the prescriptive-based design, in which fire safety designers need to follow a detailed set of rules and regulations. Using a metaphor, the two approaches can be compared to two alternative ways a chef can prepare a meal. Using a prescriptive-based approach, the chef uses a cookbook in which the exact ingredients, procedures, and tools to be used are described in detail in order to prepare a tasty meal. The chef who uses a performance-based approach can instead use any ingredient or procedure as long as the final result is a tasty dish.

The Performance based design approach makes use of Fire Safety Engineering to provide an adequate level of safety. The main scope of Fire Safety Engineering is to ensure a tenable environment during the time needed by people to evacuate [5]. This is calculated comparing the time in which the conditions of the environment are tenable (also called ASET, Available Safe Egress Time), and the time actually employed by people to reach a safe place (also called RSET, Required Safe Egress Time). The relation between ASET and RSET helps the fire safety designer understand the current level of safety of the building under consideration (i.e. if ASET < RSET the building is not safe). In addition, Fire Safety Engineering can be used to perform tenability assessment during the entire passage of time of an evacuation.

The calculation of the ASET can be made using different methods, ranging from mathematical formulae to Computational Fluid Dynamics [6]. Similarly, the calculation of the RSET can be made using different levels of sophistications [7].

1.1.1.1 History of Evacuation Modelling

In order to understand the development of evacuation modelling, an overview on the history of the study of human behaviour in fire and evacuation is needed [8]. The first generation of researchers in the field of evacuation produced a set of seminal works on human behaviour in fire and pedestrian movement and modelling ([66]; [3, 4, 9–11]). Those works carried out between the 70s and the 80s still represent the main core of the literature on which the fundaments of evacuation modelling is based. Nevertheless, there was a need for further integration of the theories developed and used to explain different issues associated with human behaviour and the calculation methods proposed for engineering applications.

Traditionally, the calculation of the RSET was in fact made using simple equations [9] based on the concepts that the movement of humans can be approximated to the movement of fluids. This idea is supported by the concept that human movement can be seen as a fluid in which each individual represents a particle. Observing a top-view of a large-scale evacuation such as the outflow of people from a stadium or a metro station, it is easy to understand that such assumption is based on the people-to-fluid similarity. In addition, the main factors taken into account during evacuation calculations were the distance to the exit, travel speeds and the flows through corridors, doors and stairs [12]. In fact, one of the main concepts proposed was the use of effective width, i.e., during evacuation movement, people use only a part of the space due to boundary layers [10].

This approach of representing human movement using hydraulic models is available in the fire safety literature [13] and it still represents a portion of the techniques used today to represent people movement.

The main issue associated with the calculation of RSET based on only hydraulic calculations is the limited representation of the impact of human factors on evacuation. One simple example is that the time needed to evacuate a building is not only the time needed to walk by the evacuees to reach a safe place from their initial location. In contrast, it includes additional times needed by the evacuees to make a decision to start their movement towards and exit or a safe place. The complexity of representing people's decision-making process led researchers to gradually develop more sophisticated modelling methods to simulate human behaviour and coupling them with the engineering approach. To address this issue, the first computer models for evacuation simulation have been developed during the 80s [14–16].

Evacuation modelling evolved in the 90s towards a gradual implementation of behavioural factors within models. This is reflected in a gradual transition between pure equation-based models (i.e. hand calculations) towards agent-based models (i.e., models based on a set of rules, such as behavioural rules, agent-to-agent and agent-to-environment interactions, etc.). To date, model developers have explored and used many techniques and behavioural theories, with regards to both the resolution of the geometry as well as the modelling methods [7]. Nevertheless, to date some of the behaviours that may occur are not completely understood and there is no agreement on the theories to be used to explain certain phenomena.

1.1.1.2 The Time-Line Model

Given the historical background, most evacuation models are based on a simple time-line model, which represent people movement as a summation of times in which the behavioural complexity is represented through agent-to-agent and agent-to-environment interactions. The increased level of sophistication leads to the need of computers to represent those complex behaviours within the time-line model. The evacuation modelling field as such is therefore born due to the need for computational modelling to solve the equations for representing behaviour as well as simulating the rules which affects their interactions with the environment and between each other.

Different models are today available for the representation of human behaviour in fire. Two reviews [12, 17] have listed and discussed the most important of them, namely the 1) behaviour sequence model [4], 2) the role-rule model [4, 18], 3) the affiliative model [3] and 4) social influence [19, 20]. Those models are theoretical frameworks which are useful to qualitatively interpret the behavioural fundaments of people decision making during fire evacuations. Nevertheless, from an engineering perspective, they are difficult to apply in a modelling framework in such a way that quantitative results (i.e. evacuation times, flows, densities, etc.) can be easily obtained. For this reason, the model that is commonly applied within evacuation models is a time-line model in which the course of events is described as a list of consecutive steps (see Fig. 1.1). This model has been included in international legislations and handbook, e.g., ISO TR16738 [21], BS PD 7974-6, [22], the Society of Fire Protection Engineering Handbook [23] etc. Each step represents a different phase of the evacuation process and the representation of people's behaviour is made through the summation of different times. Although the time-line model may be considered as an over-simplification of the behaviours that may occur, it presents the great advantage to enable a quantitative analysis of the evacuation process in a relatively short time as well as an easier implementation in evacuation models. The time-line model calculates the RSET as a sum of four times:

$$RSET = \Delta t_{\det} + \Delta t_{warn} + \Delta t_{pre} + \Delta t_{trav} \tag{1.1}$$

Where:

Δt_{det} is the detection time. This is the time from fire ignition to first people detection.

Δt_{warn} is the alarm time. This is the time from detection to a general alarm.

Δt_{pre} is the pre-evacuation time. This time includes two phases, namely the recognition time Δt_{rec} and response time Δt_{resp}. The recognition time is the period after an alarm is evident, but before people begin to respond. The response time is the period after people recognize the alarm cues and begin to respond to them, but before starting the travel phase.

Δt_{trav} is the travel time, i.e., the time required for evacuees to walk to a safe place. A safe place can be an exit or an escape route. This time also includes the time required to flow through exits and escape routes.

The evacuation time Δt_{evac} is the sum of the pre-evacuation time Δt_{pre} and the travel time Δt_{trav}. The available margin of safety can be calculated as the difference between ASET and RSET (see Fig. 1.1).

The time-line model permits the representation of an evacuation scenario as a sum of different times constituting a consecutive sequence of events. It is important to note that evacuation models are generally able to calculate the evacuation time Δt_{evac} rather than simulating the entire time-line [24]. In fact, evacuation modelling tools are focused on the calculation of evacuation times, intended as the sum of pre-evacuation and travel times, here described as pre-evacuation and movement phase.

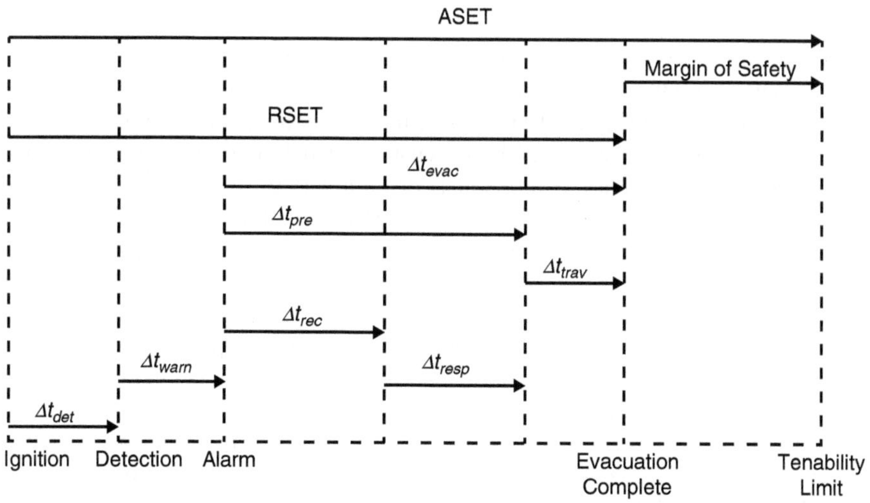

Fig. 1.1 Egress time-line model

1.1.1.3 Pre-evacuation

Different definitions are available in the literature for the pre-evacuation phase. It should be noted that different terminology is employed in evacuation modelling to refer to the time(s) to be simulated during the pre-evacuation phase. Examples include pre-travel activity time, pre-movement time and pre-evacuation time [25]. Criticism has been raised around the word pre-movement since some authors argue that people do not remain in place during this phase since they actually move in search of information [26]. The term pre-evacuation time is among the most used in evacuation modelling literature [27, 28]. This term is intended to include the variety of activities that evacuees may engage prior purposive evacuation movement. Those activities represent a delay to movement to safety (that is why this phase is sometimes also indicated in evacuation models as delay time, waiting time or response time). Different activities may take place, such as information gathering, preparation for evacuation associated with gathering personal belongings, assistance to other evacuees, alerting other people, fighting the fire, etc. [29]. It should be noted that in reality some of these activities may also take place during the movement phase, but given the additive nature of the time-line model, they are generally all included in the pre-evacuation time.

The pre-evacuation time has been defined in the literature in different ways. In the present work, we refer to the following definition: "Time taken by occupants for decision making to respond to a fire incident" [3, 23].

It should be noted that pre-evacuation times can depend on several internal and external factors, such as the level of perceived urgency [30], personal and cultural background [31], past fire experiences, training level, type of installations available in the building, emergency signage and way-finding installations [32], etc.

Evacuation models use different approaches for the simulation of the pre-evacuation time [33]. The most common method consists on the assignment of a single delay/waiting time (generally using a fixed value or a pseudo-casual number obtained from a distribution) to different individuals or categories of people [34–36]. Evacuees are therefore waiting in their initial location until their delay time is elapsed, and then they start their purposive movement towards a safe place.

An alternative approach used by evacuation models is the assignment of two single delay/waiting times to individuals or groups (representing recognition and response) [37]. Also in this case, the values can be fixed or obtained using a pseudo-random generator from pre-defined distributions. The most complex approach is the assignment of a specific behavioural itinerary, which may correspond to a set of actions that an individual or a group will perform [23, 38, 39]. This approach is used to simulate a more complex behavioural response during the pre-evacuation phase, but it often relies on the input calibration effort performed by the model user. The estimation of the pre-evacuation time to be inserted as an input in an evacuation model generally relies on the review of the literature currently available, including experimental work, case studies and previous modelling research. It should be noted that the selection of the appropriate values for pre-evacuation time depends on several factors, such as the type of population involved, the building type, people alertness, the level of familiarity with the building, etc. [40]. Evacuation models permits the use of different types of distributions such as uniform, normal, log-normal, truncated normal, gamma, beta, triangular, Weibull, Exponential, Gumbel [34–39, 41]. Different studies observed that the pre-evacuation times can be reasonably approximated in many scenarios using normal or log-normal distributions [40]. The choice of model users in terms of the distributions adopted for the representation of the pre-evacuation times relies therefore on the availability of data for that particular type of building and population under consideration as well as the assumptions adopted during the modelling phase.

1.1.1.4 Travel Time

The second component of the evacuation time is the travel time. Also in this case, different terminology is used in the literature to discuss travel time, including other terms used inter-changeably such as movement time, and walking time [42]. In the present work, the term travel time is used to indicate the movement phase, or the time required for evacuees to walk to a safe place. A safe place is not necessarily a door which leads out of a building, but it may indicate any other egress component which is considered safe in the fire safety design (e.g. a refuge area, a safe compartment, etc.). The majority of evacuation models generally terminate automatically the simulation once everybody reaches a pre-defined target safe place. The user needs to specify those targets when the model inputs are specified.

Different issues need to be taken into account during the simulation of the travel time. The travel time depends on agents' travel speed during the passage of time and the path that they walk. The main performance behavioural elements which affect travel time can be categorized as 1) travel speed, 2) route availability and usage/choice and 3) flow conditions/constraints [43].

Travel speed is the maximum uncongested travel speed at which people move towards a safe place. Unimpeded travel speeds depend on the characteristics of people under consideration (e.g. age, gender, general fitness, presence of disabilities, etc.). In this context, evacuation model often generates travel speeds for different people types as a deterministic value or in accordance to distributions (generally truncated normal distributions), i.e. similarly to pre-evacuation times. The user can define different distributions of unimpeded travel speeds and a pseudo random generator is used by the model to assign a constant unimpeded initial travel speed to each agent. Pre-set defaults of unimpeded agent travel speeds may or not already be available in the model, with maximum speeds in the order of approximately 1.8 m/s and average values for able bodied adults of about 1.0 m/s [44].

The unimpeded travel speeds are then modified at each time-step in relation to different factors. Those factors can be physical, psychological or sociological. The main physical factors include crowding (intended as the presence of other people as a physical obstacle to people movement), the type of egress component, (i.e. a stair, a corridor, an escalator, etc.), and the effects of smoke (e.g., the lower the visibility, the slower the speed [45, 46]). It is indeed common that evacuation models attempt to represent automatically the impact of proximity to others on travel speeds, generating updated pedestrian crowding and flows accordingly [7]. Similarly, different correlations today are implemented by evacuation models for the representation of the impact of smoke on travel speeds in relation to visibility [47]. Psychological factors include the representation of risk perception and motivation. Nevertheless, current evacuation models do not generally represent the impact of internal psychological factors on travel speeds. This is because they are generally based on the assumptions of self-driven particle models [48], i.e. individual unimpeded travel speeds are represented as a constant value drawn from a distribution (which may be affected by random perturbations). Sociological factors include the impact of social interactions on travel speeds. For instance group behaviours may be represented in evacuation models and a group of people may modify their speed to walk all at the same speed.

The second factor determining travel times is route availability and usage/choice. The path adopted by people can indeed depend on the routes available for evacuation and the choice (at an individual or a group level) adopted by people in terms of the route to adopt towards a safe place. Evacuation models may or not allow dynamic route availability and they include a variety of algorithms for route choice and route decision-making. Evacuation models generally allow the user to assign deterministically the routes of the users or they implement (deterministic or probabilistic) algorithms which automatically calculate the route of each agent. Those algorithms can be based on different conceptual principles, such as aiming at the representation of the shortest routes, quickest routes, the impact of familiarity to the exits, route choice in relation to available egress capacity, etc. or include a more complex agent route choice based on different conditions [7, 43].

The last factor affecting travel time is the flow conditions/constraints. Flows conditions are simulated in evacuation models in order to represent realistic number of people that can pass by a certain point over a certain period of time in relation to congestion. Various methods are available in evacuation models to represent flow conditions/constraints in different egress components (e.g. flow through a door or a corridor). For instance, the concept of effective width is widely used to calculate a specific flow rate through different egress components [10, 13]. Another approach uses experimental data to automatically calculate flows in relation to travel speeds and pedestrian densities [49]. Evacuation models may allow different degrees of user control on the calculated flows, including for instance the possibility to deterministically restrict the flow through a certain egress component.

It should be noted that depending on the type of model under consideration, and the factors included in the calculation of the travel time, different inputs and modelling assumptions may be required to the user.

1.2 Types of Models

Evacuation can be represented in computer models in a multitude of different ways, each having its own advantages and limitations. In order to understand these advantages and limitations, the model user needs to understand how the model represents both the evacuees and the geometry, e.g., the building. In addition, there can be other aspects of the model representation, such as the degree of behaviour modelling, which are important for the user to be aware of.

In subsequent sections, some of the model features commonly used to classify evacuation models are presented, and the limitations and advantages of the different types of models are discussed. Although many of the most important features are discussed below, the reader is referred to Kuligowski et al. [7] for a more comprehensive classification system.

1.2.1 Representation of Evacuees

There are two fundamentally different ways of representing people in evacuation models. In this section, these two approaches are called macroscopic and microscopic models. Each approach is associated with its own distinct advantages and disadvantages.

According to the macroscopic approach, evacuees are not represented individually in the model, but an analogy to fluid flow is instead used. The evacuees are typically represented by a two-dimensional artificial fluid, which flows through the geometry. This fluid has a specific density, which corresponds to people density, and a velocity.

One of the first macroscopic approaches for representation of evacuees was presented by Predtechenskii and Milinskii [11]. According to Predtechenskii and Milinskii, evacuation can be treated as a two-dimensional flow problem. The evacuees are represented by a fluid with a dimensionless density. The calculation of dimensionless density, which is the total area of the evacuees divided by the total area of the space, is illustrated in Fig. 1.2.

In order to get a realistic representation of crowd movement, the macroscopic approach relies on correlations between travel speed and density. This type of correlation has been presented by, for example, Predtechenskii and Milinskii [11] for different geometric configurations. Examples of correlations for flat surfaces, openings, stairs up and stairs down are shown in Fig. 1.3.

When modelling movement with macroscopic representation, evacuees are first transformed in the computer model into a fluid with a specific density. This fluid then 'flows' out from the building. At locations where the building geometry changes the density of the flowing fluid is changed. For example, a flow restriction, e.g., a corridor that becomes narrower, will lead to compression, i.e., a higher density, which will in turn influence the velocity according to the previously mentioned correlations (see Fig. 1.3). Today, there are relatively few evacuation models that use the macroscopic representation of evacuees [50].

The microscopic approach is today the most common way of representing evacuees in computer models [51]. Each evacuee is represented by an autonomous agent with certain properties, e.g., pre-movement time, travel speed, etc. A crowd in hence built up of a group of agents acting together in the model. This approach is arguably more similar to reality than the macroscopic representation because crowd moment is, in reality, the result of individual movement of people in larger groups [52].

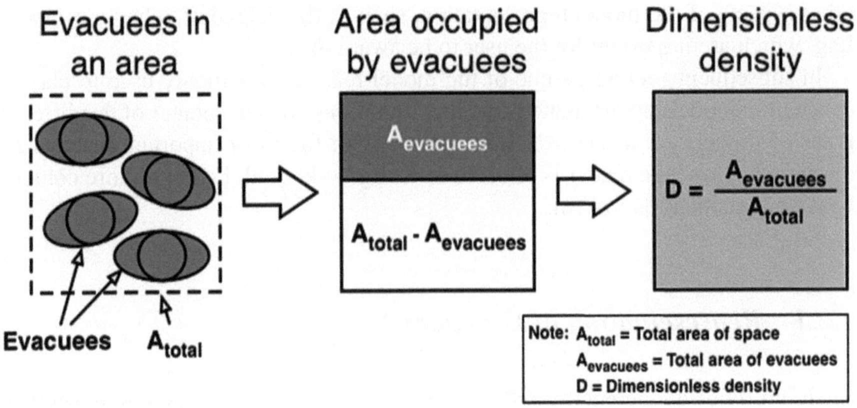

Fig. 1.2 Representation of evacuees as an area with a dimensionless density [11]

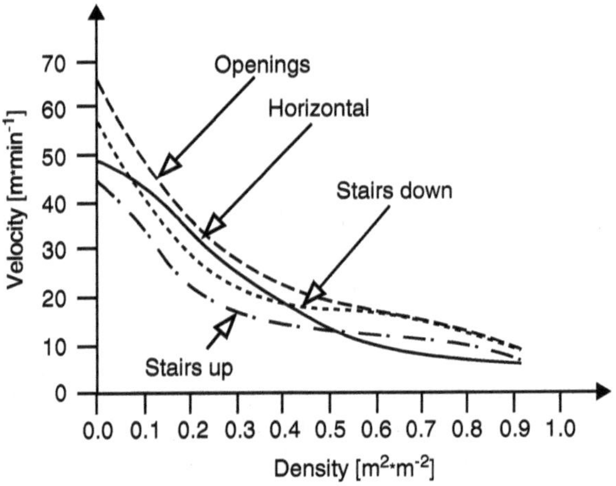

Fig. 1.3 Correlations between travel speed and dimensionless density for different geometric configurations [11]

Agents act according to behavioural rules defined in the evacuation model. These rules can relate to agent-to-agent or agent-to-environment interactions. An example of an agent-to-agent rule that is used in some models is that agents will reduce their travel speed as a function of the distance to persons in front, i.e., the inter-person distance [35]. The exact rules used in any computer model depend on the specifics of the model, e.g., how the geometry is represented, etc. However, a common feature is that the behavioural rules typically generate emergent phenomena [52], i.e., the interactions between agents and agents/environment give rise to complex phenomena on the macroscopic level. Microscopic representation in evacuation models can hence be seen as a bottom-up approach.

Since most evacuation models today use the microscopic rather than the macroscopic approach [51] the description below will be based on the assumption that evacuees are represented by agents in models. However, much of the discussions can still be generalised to macroscopic models.

1.2.2 Representation of Geometry

There are three basic ways to represent the geometry, i.e., the building, in evacuation models. In this section, these three approaches are called course network, fine network and continuous models. Each approach is associated with its own distinct advantages and disadvantages, which will be explained more thoroughly below.

In course network models, a network with nodes, which are connected with arcs, represents the geometry (see Fig. 1.4). A node is typically a representation of an area inside the geometry, e.g., a room of a building or a corridor, but can also represent connection between areas, e.g., a door opening. In the latter case, a more exact representation of the walking distances along the arcs can sometimes be achieved (see left image in Fig. 1.4).

In some cases, it might be necessary to use several nodes to represent an area, i.e., in the case of very long corridors. In addition, one or more nodes are assigned as safe places, i.e., the nodes that evacuees move towards when the simulation is started.

Nodes and arcs in course network models can be given specific characteristics in order to get more realistic representation of the geometry. Nodes can be given a capacity, e.g., a maximum number of agents allowed in the node, in order to restrict the number of people occupying a certain area of the geometry, e.g., a room. Specifying a node capacity can be essential in order to avoid unrealistic overcrowding. Arcs can instead be assigned different types of flow restrictions. Limiting the number of agents travelling along an arc or the number of agents entering an arc per simulation time step are different ways of restricting the flow. In addition, travel speed along an arc is typically set in the program, which determines the time taken to move from node to node.

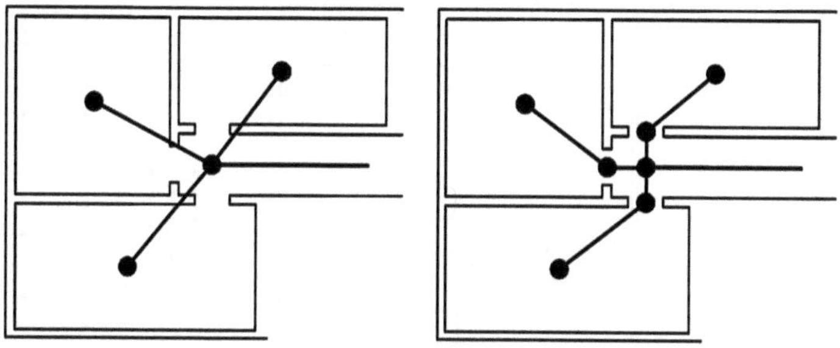

Fig. 1.4 Schematic representations of course network models with nodes only in areas (*left*) versus areas and connections (*right*)

A common feature for all course network models is that a relatively simple network with nodes and arcs represents the geometry. This simplification can potentially influence the validity of the simulation results. It is therefore essential that the expected movement paths of evacuees are represented correctly and that no paths are forgotten when networks are created. Inaccurate or too simplified representation of the geometry will lead to misleading simulation results. The simplicity of the geometry representation is, however, also an advantage since simulation times are typically very short. This means that a large number of scenarios can easily be simulated, which makes it feasible to perform sensitivity analyses even for simulations involving large numbers of agents.

Because of the simplified representation of the geometry in course network models, it often not possible to get detailed information about such aspects as the exact location of agents or crowding conditions. The resolution of simulation results is only as accurate as the network representing the geometry. It is, for example, not possible to get the exact location of an agent within a specific room because a room is typically only represented by a node containing a certain number of agents. Similarly, it is not possible to accurately simulate the build-up of over-crowding as the nodes can only be filled to the maximum capacity specified by the user.

Although it might be possible to identify areas of the geometry where nodes are filled to maximum capacity, it is not possible to identify dangerous over-crowding within the separate nodes, e.g., in specific rooms.

When evacuation models were first introduced in the 1980s, the course network models were the most common type of model. Today, the course network models represent a small part of available evacuation models [51].

Fine network models represent the geometry by either a grid with cells or a network with nodes connected by arcs (see Fig. 1.5). In the case of grids, different shapes of the cells can be chosen, e.g., square or hexagonal (see Fig. 1.6). Agents move from cell to cell in the simulations, and each cell can be occupied by a maximum of one agent. Obstacles, e.g., walls and pillars, are typically represented

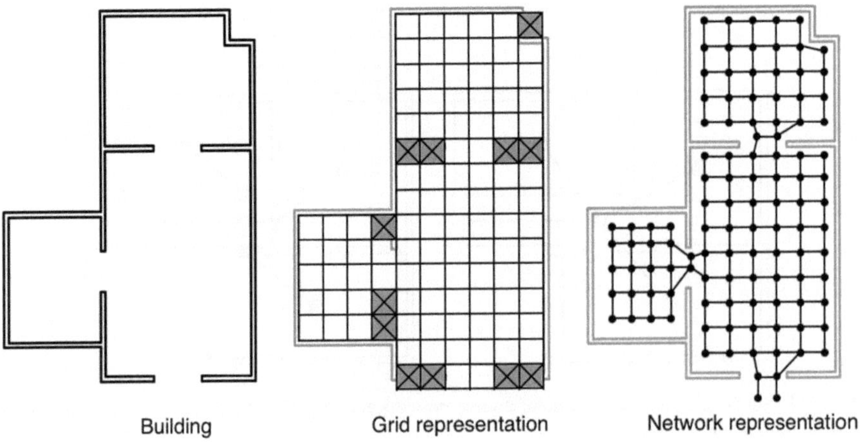

Fig. 1.5 Schematic representations of fine network models with grids and networks

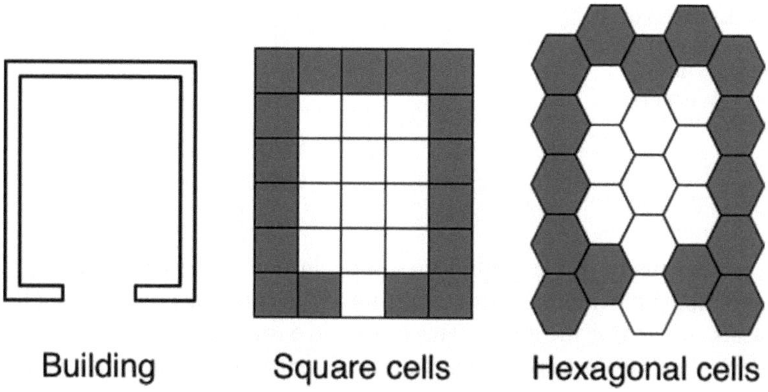

Fig. 1.6 Different shapes of cell used in fine network models together with the implementation of obstacles

by blocked cells, i.e., cells that cannot be occupied by agents. This means that a blockage needs to be adjusted to the grid, which can result in different implementations of the geometry depending on the configuration of the grid (see Fig. 1.6).

In the case of networks, each node can typically be occupied by only one agent. These agents can move from node to node via arcs. This approach does not require any blocked nodes, but instead a blockage, e.g. wall, is represented by the lack of arcs connecting nodes on opposite sides of the blockage (see Fig. 1.5). If networks instead of grids are used it is often possible to represent the geometry more accurately in the model.

One advantage of fine network models is that the structured representation of the geometry, i.e., a grid or network, ensures that simulations typically run very smoothly. For example, counter-flow situations are typically easier to implement in fine network models than in a continuous models. The structured representation of the geometry usually also means that simulations with large number of agents are possible and do not require extreme computational resources.

The assumption that each cell or node can only contain one agent can be seen as a limitation of fine network models. Because the cell size (or distance between nodes) is typically fixed throughout a simulation, the maximum achievable density is also pre-defined. For example, if a fine network model uses a cell size of 0.5 by 0.5 m the maximum number agents per square meter, i.e., the maximum density, will be 4 persons/m^2. This value might be lower than densities observed in the field [11], but given the assumptions in the model these situations are not possible to simulate. Similarly, the way the geometry is represented may result in difficulties to correctly implement the influence of density on travel speed. There are currently many fine network models available on the market [38, 39].

In continuous models, the geometry is represented by a continuous space containing agents. These agents have coordinates describing their location as well as a body shape and size. Examples of commonly used approximation of human body shape are the representation of a body circle, or a torso and two shoulder circles (see Fig. 1.7) [35, 36]. Rather than being restricted by a grid or network, these agents can

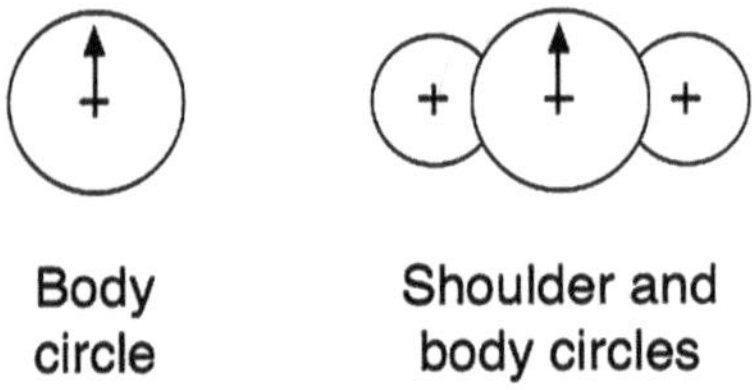

Fig. 1.7 Examples of representation of an agent in continuous models (*single circle* or *three circles*) [35, 36]

move around freely in the continuous space. In order to get realistic movement, rules describing the minimum distance from an agent to other agents and obstacles, e.g., walls, are used in the model. This is done to ensure that the bodies of agents do not overlap and that agents keep a realistic distance to obstacles. In addition, care must be taken in continuous models to use an iteration time step that is sufficiently short, since a too long time step can result in agents moving through obstacles in the simulations.

Continuous models allow the closest representation of actual people movement. In the real world, people also move in a continuous space and constantly take account of the movement of others and try to avoid collisions with obstacles.

However, the fact that the simulation process is realistic does not guarantee that continuous models are always better than other types of models, because the validity of simulation results depend on exactly how the movement process has been implemented in the model and how the model is used.

A potential disadvantage of continuous models is that they are relatively complex, which implies that calculation times can be much longer than for other types of evacuation models. A higher resolution of the geometry representation is adopted in the models, but this creates challenges with regards to the rules governing movement of agents. For example, it is relatively complicated to simulate counter-flow situations or movement at high densities. These types of phenomena can potentially be solved with sub-models, e.g., models that describe the route negotiation behaviour in counter-flow situations, but they will also add to the computation time required to simulate movement. However, the complexity can also be seen as an advantage as it allows a closer representation of reality. For example, in a real world situation, people will also negotiate the route choice with others in a counter-flow situation in order to avoid a collision.

It should be noted that recent studies introduced the concept of hybrid models in evacuation models, i.e. two or more of the previously mentioned space representation are combined in a single evacuation model [53].

1.2.3 Other model Characteristics

Although the representation of the evacuees and the geometry are the two main aspects used when categorizing evacuation models, there are a multitude of other aspects that can be used in the categorisation process. Kuligowski et. al. [7] has categorized models depending on how the behaviour of people is represented and how the movement is solved. In addition, integration with fire simulation tools and visual representation of the simulation results have been mentioned as possible aspects to include in the categorisation process [7]. However, it is argued that representation of the evacuees and the geometry are the two essential parts that always need to be known in order understand the underlying assumptions of an evacuation model.

1.3 Modelling Methods

This section discusses some of the most common assumptions adopted in evacuation models for the simulation of agent movement. In particular, since the majority of the evacuation models use today are microscopic agent-based models [51], the description of the modelling methods in this section refers to hierarchies of behaviours adopted for rule-based simulations of individual agents.

Different hierarchies of behaviours can be employed to discuss the simulation of the evacuation process. This means that there may be different levels of interactions between each individual agent and the environment. In this section, we refer to the sub-division made by Reynolds [54], which can be applied for most evacuation models in which the agents are considered as autonomous entities (microscopic models). Nevertheless, some of the concepts presented here can be extended for other types of simulation tools (e.g., macroscopic models). The reader of this section should carefully evaluate the features of the evacuation model in use and make a case-to-case evaluation on the fundamental assumptions of the model (i.e. microscopic vs macroscopic) and assess which modelling strategies have been adopted by the developer and to which level of representation those apply.

The classification of hierarchies of behaviours [54] is used here as a starting point to understand the fundamental assumptions used by evacuation models to represent people movement. Three main levels of movement behaviours can be identified in evacuation models, namely:

1) Path-finding
   This level includes the simulation of route choice in time or space as well
2) Local Movement
   This level includes the simulation of space navigation intended as the representation of the interactions of individual agents with the environment. It is intended as the individual interactions with other agents and the general environmental conditions, such as the geometry available for navigation, hazard conditions, etc.
3) Locomotion
   This level refers to the representation of agent motion at an individual level, considering the simulation of its body movement, e.g., body sway.

In order to explain Reynolds' three levels of behaviours for the representation of agent movement, an analogy with the movement of a motorcycle driven by a person is presented. The level of path-finding corresponds to the path chosen by the driver to reach a final destination (or target) or/and the itinerary that the driver is intended to follow. The local movement level corresponds instead to the steering actions that the driver needs to do to avoid colliding with other vehicles and obstacles and remain on the road. Finally, the locomotion level is the actual movement of the motorcycle, intended as the wheel and motorcycle movement caused by the maneuvering that the driver does while navigating with the steering wheel in the road.

1.3.1 Path-Finding

The first level of behaviour includes the representation of route choice of the agents at a local or global level. This is associated with intermediate or final targets which agents want to reach and the algorithms employed to calculate their preferred route. Path-finding is usually intended as the strategic level of agent navigation, i.e., the goal(s) of the journey of evacuees. Most evacuation models generally assume that an agent "knows" from the beginning of the simulation what its target(s) will be. Some exceptions exist, e.g., evacuation models may rely on visual perception for path-finding [55] rather than navigating towards a known target exit.

Routes may be calculated in accordance to different criteria such as 1) fastest route, 2) shortest route, 3) user defined [7]. Different criteria and algorithms may be used to represent path-finding at a global level (the route to the final exit to be reached) and at a local level (if there are intermediate targets or if different algorithms are used for re-routing under certain local conditions).

The path-finding algorithms may vary substantially in relation to the type of space representation. For instance, in case of coarse network models, well known methods from the field of graph search algorithms, e.g. the A* algorithm [56] or the Dijkstra's algorithm [57], can be used in pedestrian navigation to search the shortest route (see Fig. 1.8a). Nevertheless, those algorithms are not very common in current evacuation models given the fact that they can be used in case of a coarse representation of the geometry (which is today not as frequent as the other types of representations [51]). More common algorithms employed for path-finding in evacuation models rely on the concept of distance maps [35, 39]. A distance map represents the space using numerical values assigned to different sub-areas in line with the total travel distance needed to reach an exit from that part of the building (see Fig. 1.8b).

An alternative approach for path-finding (different than the calculation of a shortest or fastest route) is the use of vector fields to represent the preferred directions of the agents over time during space navigation (see Fig. 1.8c). The vector fields can be calculated in different manners. An interesting example of such method is the use of an approximate solution to a potential flow problem of a two-dimensional incompressible fluid where the target exit acts as a fan extracting the fluid from the domain [37]. In case of more than one available exit, it would then be possible to calculate multiple vector fields and then provide agent assignment to specified routes through specific vector fields.

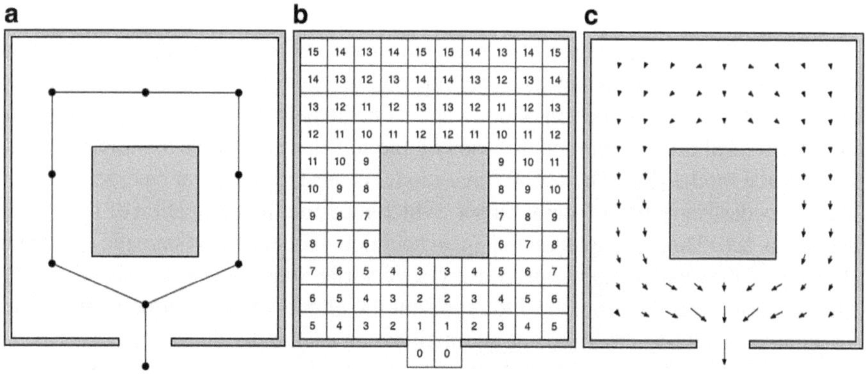

Fig. 1.8 Examples of path-finding algorithms (**a**) is an exemplary graph-based solution for path-finding; (**b**) is an exemplary solution of a distance map solution for path-finding; (**c**) is an exemplary solution of a vector field solution for path-finding

Route choice can also be modified in accordance with different conditions, such as congestion levels, smoke, group interactions, etc. Evacuation models can therefore allow modifying the target exits during the passage of time.

1.3.2 Local Movement

Local movement algorithms deal with the representation of the agent-to-agent and agent-to-environment interactions at a local level. The use of local movement representation assumes that agents are simulated as autonomous entities (i.e., a microscopic modelling approach is used). Local movement is mainly relevant for continuous models since in coarse and fine network models, people movement is solved at a global level (i.e., path-finding is used also to solve local conflict resolution). In some cases, fine network models may implement additional algorithms to solve local movement [39].

Many different algorithms are used in evacuation modelling to solve local movement and agent-to-agent and agent-to-environment interactions or additional environmental conditions such as hazards, etc. The local movement algorithms can be inspired by different conceptual assumptions. In this part, some of the most used assumptions and local movement algorithms are discussed.

One of the most used assumptions to simulate local movement is the concept of steering behaviour of autonomous (i.e., individual) agents. The physics of the agent local movement are based on behaviourally determined steering forces which are applied to the mass of each agent at each time-step [54]. Those steering forces are generally solved using agent-based rules (such as "separation", "cohesion", "alignment", etc.). This model was first developed to simulate the movement of birds in computer animation, e.g., the model Boids [58] as a result of the interactions within a particle system.

A variation of the classical steering model is the simulation of agents based on self-driven particles (i.e., called self-driven or self-propelled particle models). Each agent walks in this case at a constant speed but respond to random perturbations by adopting the average direction of motion of other agents in the neighbourhood [48]. One of the most used local movement models based on the self-driven particle assumption is the social force model [59]. The social force model is used to represent emergent behaviours of pedestrians using "social forces" which measure the motivation of individual agents to act. This is implemented in evacuation models by systematic temporal changes of the desired agent speed using a vectorial quantity (i.e. the social force). The social force is therefore used to represent the effects of the environment (agent-to-agent and agent-to-environment interactions) on each individual agent.

Completely different methodological assumptions can be used to simulate the local movement of the agents. An example is the use of inter-person distance, intended as the distance between the bodies of two agents [35]. In this case, the actual travel speed of individual agents is an emergent property that is affected by the proximity of other agents or obstacles (that is an indirect measure of densities). The overall speed of a crowd would then decrease as the distance between agents decrease.

1.3.3 Locomotion

The locomotion level is the highest level of resolution for the representation of the movement of the agents. It represents the conversion of the local movement rules into the motion of the agent "body". This can also be defined as the transmission of control signals into motion [54]. It is important to note that evacuation models often do not characterize locomotion and (if present) the locomotion schemes are generally independent of the local movement methodology employed. In other words, locomotion is often a pure visualization feature in evacuation models rather than a sophisticated scheme used to represent movement and navigation also at a high level of resolution.

1.4 Limitations of Models

The analysis and use of the results of evacuation models requires the user to have a deep knowledge about their limitations. Two main issues are discussed in this section in order to provide guidance to a model user approaching an evacuation model simulation. The first issue is associated with the need to assess the limitations of the underlying assumptions on which the predictive capabilities of a model are based as well as the degree of user control on them. To study this issue, it is necessary to discuss modelling assumptions in light of the current knowledge on human behaviour in fire as well as the data-sets and default settings embedded in the model and their impact on results. The second important limitation discussed in this section is the uncertainty associated with evacuation model predictions. To discuss this issue, uncertainties in evacuation modelling have been classified and explained.

1.4.1 Model Capabilities

The first analysis that an evacuation model user is required to do to assess model limitations is a detailed study of the model features and the associated predictive capabilities. Different reviews are today available [7, 51, 60, 61] which allow model users to identify the features and sub-models included in evacuation models. This preliminary analysis permits identifying if a model is able to represent a specific set of behaviours and the type of sub-model/assumptions employed to achieve this scope. Given the current knowledge on human behaviour in fire, there may be cases in which there is no agreement in the evacuation simulation literature between the most appropriate modelling assumptions to represent a specific behaviour. Consequently, model users may be in a difficult position of choosing among models and evaluate their predictive capabilities which are based on different assumptions, and for which there is no agreed "best" modelling method. Nevertheless, an expert user can identify if a specific set of assumptions may be more or less suitable for different types of evacuation model applications. This is associated with the fact that certain sub-models or methods may be able to better predict certain behaviour rather than others. The assessment of a suitability of a model is easier in the cases in which certain models do not include sub-models to simulate specific egress components or behaviours (e.g., a model which does not include a sub-model for lift evacuation should not be used if the evacuation strategy of a building includes this egress component). In fact, it is important to assess to which extent model results are dependent on embedded modelling assumptions and if and how they can be affected by the user.

Evacuation model predictive capabilities may be strongly dependent on both the user's model input assignment as well as model developer assumptions. It is important to note that at the present stage of research, the validity of the results of evacuation models heavily relies on the availability of experimental data for model input calibration, given the lack of an agreed comprehensive predictive theory of human behaviour in fire [29]. An additional complication to this issue, is the fact that models may include different types of default settings and different types of user's control on them [44].

Three main categories of default settings are today available in evacuation models [44, 62], namely 1) no default settings, 2) a single default setting, 3) multiple pre-defined settings. When no default settings are embedded in the model, the user is required to configure the input using a full range of data-sets to run the model. When there is a single "factory" setting, the user can speed up the process of configuring the input by directly applying the default setting available in the model. In case of multiple pre-defined settings, the user has an initial set of possible libraries or settings, among which he/she can choose.

The user control on default settings may substantially vary among models. Model users may for instance be forced to use a certain data-set even outside is original range of applications [63]. It is important to note that transparency plays a key role for understanding the underlying settings and embedded data-sets behind a model. It may occur that the basic assumptions of a model are not fully available to the user. This is generally associated with the type of model under consideration, i.e., whether the model is commercial or not, open source projects, etc. and the documentation available with the model.

1.4.2 Uncertainties in Evacuation Modelling

Another important limitation of evacuation models is linked to the difficulties in estimating the uncertainty associated with the predicted results. In this context, the interpretation of the results obtained using evacuation modelling tools need to be done assessing all uncertainties associated with their predictive capabilities. Similarly to other predictive simulation tools, e.g., fire models, evacuation models are affected by different types of uncertainties.

In evacuation modelling, four different types of uncertainties should be taken into consideration for the analysis of the results, namely 1) measurement uncertainty, 2) model input uncertainty, 3) intrinsic uncertainty and 4) behavioural uncertainty [64]. The classification of those uncertainties is based on the analysis of uncertainties observed in fire modelling [65], with the exception of behavioural uncertainty, which is typical of evacuation modelling [63, 64]. Measurement uncertainty refers to the uncertainty linked with the experimental measurement itself. Model input uncertainty depends on the parameters, distributions, etc. used as input for the model. Intrinsic uncertainty depends on the assumptions of the model (i.e. physical, mathematical and modelling assumptions) and he methods used in the formulation of the model. Behavioural uncertainty is linked with the use of stochastic variables or distributions employed in evacuation modelling to represent human behaviour. Uncertainties are further discussed in the chapter of the present book on model validation (see Chap. 5).

It should be noted that on top of the uncertainties associated with the simulation of a single evacuation scenario, the uncertainties associated with the definition of the scenario itself should also be taken into consideration during the use of evacuation models for a risk analysis. In fact, user's assumptions and uncertainty in terms of assumed occupant load, initial locations of the agents, fire conditions affecting evacuation behaviours, etc. should be taken into consideration during the analysis use of evacuation model results.

References

1. Fahy, R. F., Proulx, G., & Aiman, L. (2012). Panic or not in fire: Clarifying the misconception. *Fire and Materials, 36*, 328–338. doi:10.1002/fam.1083.
2. Quarantelli, E. L. (1954). The nature and conditions of panic. *American Journal of Sociology, 60*, 267–275.

3. Sime, J. D. (1985). Movement toward the familiar: Person and place affiliation in a fire entrapment setting. *Environment and Behavior, 17*, 697–724. doi:10.1177/0013916585176003.
4. Canter, D. V. (1990). *Fires and human behaviour*. London: Fulton.
5. Purser, D. (2003). ASET and RSET: Addressing some issues in relation to occupant behaviour and tenability. *Fire Safety Science, 7*, 91–102. doi:10.3801/IAFSS.FSS.7-91.
6. Olenick, S. M. (2003). An updated international survey of computer models for fire and smoke. *Journal of Fire Protection Engineering, 13*, 87–110. doi:10.1177/1042391503013002001.
7. Kuligowski, E. D., Peacock, R. D., & Hoskins, B. L. (2010). *A review of building evacuation models* (2nd ed.) (Tech. Note 1680). Washington, DC: NIST.
8. Bryan, J. L. (2002). A selected historical review of human behavior in fire. *Fire Protection Engineering, 16*, 4–10.
9. Pauls, J. L. (1980). Building evacuation: Research findings and recommendations. In D. Canter (Ed.), *Fires and human behaviour* (pp. 251–275). New York: Wiley.
10. Pauls, J. L. (1980). *Effective width model for evacuation flow in buildings*. Presented at the Engineering Applications of Fire Technology (pp. 215–232). Washington, DC: National Bureau of Standards.
11. Predtechenskii, V. M., & Milinskii, A. I. (1978). *Planning for foot traffic flow in buildings*. New Delhi: Amerind Publishing.
12. Kobes, M., Helsloot, I., de Vries, B., & Post, J. G. (2010). Building safety and human behaviour in fire: A literature review. *Fire Safety Journal, 45*, 1–11. doi:10.1016/j.firesaf.2009.08.005.
13. Gwynne, S. M. V., & Rosenbaum, E. (2008). Employing the hydraulic model in assessing emergency movement. In P. J. Di Nenno (Ed.), *SFPE Handbook of fire protection engineering*. Quincy, MA: National Fire Protection Association.
14. Kisko, T. M., & Francis, R. L. (1985). EVACNET+: A computer program to determine optimal building evacuation plans. *Fire Safety Journal, 9*, 211–220. doi:10.1016/0379-7112(85)90009-8.
15. Takahashi, K., Tanaka, T., & Kose, S. (1989). An evacuation model for use in fire safety design of buildings. *Fire Safety Science, 2*, 551–560. doi:10.3801/IAFSS.FSS.2-551.
16. Watts, J. M. (1987). Computer models for evacuation analysis. *Fire Safety Journal, 12*, 237–245. doi:10.1016/0379-7112(87)90008-7.
17. Fridolf, K., Nilsson, D., & Frantzich, H. (2011). Fire evacuation in underground transportation systems: A review of accidents and empirical research. *Fire Technology, 49*, 451–475. doi:10.1007/s10694-011-0217-x.
18. Tong, D., & Canter, D. (1985). The decision to evacuate: a study of the motivations which contribute to evacuation in the event of fire. *Fire Safety Journal, 9*, 257–265. doi:10.1016/0379-7112(85)90036-0.
19. Deutsch, M., & Gerard, H. B. (1955). A study of normative and informational social influences upon individual judgment. *Journal of Abnormal and Social Psychology, 51*, 629–636. doi:10.1037/h0046408.
20. Latané, B., & Darley, J. M. (1970). *The unresponsive bystander: Why doesn't he help?* New York: Appleton-Century Crofts.
21. International Standards Organization. (2009). *ISO/TR 16738:2009 Fire-safety engineering—Technical information on methods for evaluating behaviour and movement of people*. Geneva, Switzerland: Author.
22. British Standards Institution. (2004). *The application of fire safety engineering principles to fire safety design of buildings. Part 6: Human factors: Life safety strategies—Occupant evacuation, behaviour and condition (sub.system 6)*. London: Author.
23. Proulx, G. (2002). Movement of people: The evacuation timing. In: *SFPE handbook of fire protection engineering* P. Di Nenno (pp. 3-341–3-366, chap. 3–13). Quincy, MA: National Fire Protection Association.
24. Gwynne, S. M. V., Purser, D. A., & Boswell, D. L. (2011). *Pre-warning staff delay: A forgotten component in ASET/RSET calculations*. Presented at the Pedestrian and Evacuation Dynamics (pp. 243–253). Gaithersburg, MD: Springer.
25. Purser, D. (2008). Dependence of modelled evacuation times on key parameters and interactions. *Fire Safety Science, 9*, 353–364. doi:10.3801/IAFSS.FSS.9-353.

26. McConnell, N. C., Boyce, K. E., Shields, J., Galea, E. R., Day, R. C., & Hulse, L. M. (2010). The UK 9/11 evacuation study: Analysis of survivors' recognition and response phase in WTC1. *Fire Safety Journal, 45*, 21–34. doi:10.1016/j.firesaf.2009.09.001.
27. Kuligowski, E. D., & Mileti, D. S. (2009). Modeling pre-evacuation delay by occupants in World Trade Center Towers 1 and 2 on September 11, 2001. *Fire Safety Journal, 44*, 487–496. doi:10.1016/j.firesaf.2008.10.001.
28. Sherman, M. F., Peyrot, M., Magda, L. A., & Gershon, R. R. M. (2011). Modeling pre-evacuation delay by evacuees in World Trade Center Towers 1 and 2 on September 11, 2001: A revisit using regression analysis. *Fire Safety Journal, 46*, 414–424. doi:10.1016/j.firesaf.2011.07.001.
29. Kuligowski, E. (2011). Predicting human behavior during fires. *Fire Technology, 49*, 101–120. doi:10.1007/s10694-011-0245-6.
30. Proulx, G. (2003). Recollection, identification and perceived urgency of the temporal-three evacuation signal. *Journal of Fire Protection Engineering, 13*, 67–82. doi:10.1177/1042391503013001004.
31. Galea, E. R., Deere, S., Sharp, G., Filippidis, L., & Hulse, L. M. (2010). *Investigating the impact of culture on evacuation behaviour*. Presented at the 12th International Conference and Exhibition on Fire Science and Engineering, Interflam (pp. 879–892). Nottigham, England: Interscience Communications.
32. Nilsson, D. (2009). *Exit choice in fire emergencies: Influencing choice of exit with flashing lights*. Lund, Sweden: Department of Fire Safety Engineering and Systems Safety, Lund University.
33. Kuligowski, E. D. (2008). *Modeling human behavior during building fires* (Tech. Note No. 1619). Gaithersburg, MD: National Institute of Standards and Technology.
34. Bensilum, M., & Purser, D. (2003). Grid flow: An object-oriented building evacuation model combining pre-movement and movement behaviours for performance-based design. *Fire Safety Science, 7*, 941–952. doi:10.3801/IAFSS.FSS.7-941.
35. Thompson, P. A., & Marchant, E. W. (1995). A computer model for the evacuation of large building populations. *Fire Safety Journal, 24*, 131–148. doi:10.1016/0379-7112(95)00019-P.
36. Thunderhead Engineering. (2014). *Verification and validation—Pathfinder* 2013.1.
37. Korhonen, T., & Hostikka, S. (2009). *Fire dynamics simulator with evacuation: FDS+Evac technical reference and user's guide* (Working paper No. 119). Espoo: VTT Technical Research Center of Finland.
38. Gwynne, S., Galea, E., Lawrence, P., & Filippidis, L. (2001). Modelling occupant interaction with fire conditions using the building EXODUS evacuation model. *Fire Safety Journal, 36*, 327–357. doi:10.1016/S0379-7112(00)00060-6.
39. Mott MacDonald Simulation Group. (2012). *Simulation of transient evacuation and pedestrian movements STEPS user manual*. Croydon, England: Author.
40. Purser, D. A., & Bensilum, M. (2001). Quantification of behaviour for engineering design standards and escape time calculations. *Safety Science, 38*, 157–182. doi:10.1016/S0925-7535(00)00066-7.
41. Fahy, R. F. (2000). *New developments in EXIT89* (No. NISTIR 6588). Gaithersburg, MD: National Institute of Standards and Technology.
42. Gwynne, S. M. V. (2010). *Conventions in the collection and use of human performance data* (No. NIST GCR 10-928). Gaithersburg, MD: National Institute of Standards and Technology.
43. Gwynne, S. M. V., Kuligowski, E., Spearpoint, M., & Ronchi, E. (2013). Bounding defaults in egress models. *Fire and Materials*. doi:10.1002/fam.2212.
44. Gwynne, S. M. V., Kuligowski, E. D., & Spearpoint, M. (2012). *More thoughts on defaults*. Presented at the Fifth International Symposium on Human Behaviour in Fire. (pp. 9–23). Cambridge, England: Interscience Communications.
45. Frantzich, H., & Nilsson, D. (2003). *Utrymning genom tät rök: beteende och förflyttning* [Evacuation in dense smoke: Behaviour and movement] (Tech. Rep. 3126). Lund University.
46. Jin, T. (2008). Visibility and human behavior in fire smoke. In: *SFPE handbook of fire protection engineering* (3rd ed., pp. 2-42–2-53). Quincy, MA: National Fire Protection Association.

47. Ronchi, E., Gwynne, S. M. V., Purser, D. A., & Colonna, P. (2013). Representation of the impact of smoke on agent walking speeds in evacuation models. *Fire Technology, 49*, 411–431. doi:10.1007/s10694-012-0280-y.
48. Vicsek, T., Czirók, A., Ben-Jacob, E., Cohen, I., & Shochet, O. (1995). Novel type of phase transition in a system of self-driven particles. *Physical Review Letters, 75*, 1226–1229. doi:10.1103/PhysRevLett.75.1226.
49. Seyfried, A., Steffen, B., & Lippert, T. (2006). Basics of modelling the pedestrian flow. *Physica A, 368*, 232–238. doi:10.1016/j.physa.2005.11.052.
50. Karkin, I., Grachev, V., Skochilov, A., & Zverev, V. (2010). "FlowTech" and "EvaTech": Two computer-simulation methods for evacuation calculation. In W. W. F. Klingsch, C. Rogsch, A. Schadschneider, & M. Schreckenberg (Eds.), *Pedestrian and Evacuation Dynamics 2008* (pp. 537–545). Heidelberg: Springer.
51. Ronchi, E., & Kinsey, M. (2011). Evacuation models of the future: Insights from an online survey on user's experiences and needs. In J. Capote et al. (Ed.), *Advanced research workshop evacuation and human behaviour in emergency situations EVAC11* (pp. 145–155). University of Cantabria (Spain).
52. Nilsson, D., & Uhr, C. (2009). *Complex systems—A holistic approach for understanding and modelling fire evacuation behaviour*. Presented at the 4th International Symposium on Human Behaviour in Fire (pp. 525–530). Cambridge: Robinson College.
53. Chooramun, N. (2011). *Implementing a hybrid spatial discretisation within an agent based evacuation model*. London: University of Greenwich.
54. Reynolds, C. W. (1999). *Steering behaviors for autonomous characters*. Presented at the Game developers conference (pp. 763–782). San Francisco, CA: Miller Freeman Game Group.
55. Pan, X. (2009). *Computational modeling of human behavior for emergency egress analysis: a multi-agent based simulation approach*. Saarbrücken, Germany: Lambert Academic Publishing.
56. Hart, P., Nilsson, N., & Raphael, B. (1968). A formal basis for the heuristic determination of minimum cost paths. *IEEE Transactions on Systems, Science, and Cybernetics, 4*, 100–107. doi:10.1109/TSSC.1968.300136.
57. Misa, T. J. (2010). An interview with Edsger W. Dijkstra. *Communications of the ACM, 53*, 41. doi:10.1145/1787234.1787249.
58. Reynolds, C. W. (1987). Flocks, herds and schools: A distributed behavioral model. *ACM Computer Graphics, 21*, 25–34. doi:10.1145/37401.37406.
59. Helbing, D., & Molnár, P. (1995). Social force model for pedestrian dynamics. *Physical Review E, 51*, 4282–4286. doi:10.1103/PhysRevE.51.4282.
60. Gwynne, S., Galea, E. R., Owen, M., Lawrence, P. J., & Filippidis, L. (1999). A review of the methodologies used in the computer simulation of evacuation from the built environment. *Building and Environment, 34*, 741–749. doi:10.1016/S0360-1323(98)00057-2.
61. Ronchi, E., & Nilsson, D. (2013). Fire evacuation in high-rise buildings: a review of human behaviour and modelling research. *Fire Science Reviews, 2*, 7. doi:10.1186/2193-0414-2-7.
62. Ronchi, E., Gwynne, S., & Purser, D. A. (2011). *The impact of default settings on evacuation model results: A study of visibility conditions vs occupant walking speeds*. Presented at the Advanced Research Workshop-Evacuation and Human Behaviour in Emergency Situations. Santander, Spain: Universidad de Cantabria, GIDAI.
63. Ronchi, E., Fahy, R., Colonna, P., & Berloco, N. (2014). Validation and calibration of the EXIT89 evacuation model for road tunnel evacuation applications. In U. Weidmann, U. Kirsch, & M. Schreckenberg (Eds.), *Pedestrian and evacuation dynamics 2012* (pp. 543–550). Cham, Switzerland: Springer International Publishing.
64. Ronchi, E., Reneke, P. A., & Peacock, R. D. (2014). A method for the analysis of behavioural uncertainty in evacuation modelling. *Fire Technology, 50*, 1545–1571. doi:10.1007/s10694-013-0352-7.
65. Hamins, A., & McGrattan, K. (2007). *Verification and validation of selected fire models for nuclear power plant applications* (No. NUREG-1824). Gaithersburg, MD: National Institute of Standards and Technology.
66. Fruin, John J. (1987) *Pedestrian Planning and Design*. (Revised Edition). Elevator World, Inc, Mobile, AL.

Chapter 2
Developments in Tenability and Escape Time Assessment for Evacuation Modelling Simulations

David A. Purser

2.1 Introduction

Fire safety in any built environment depends on the integrated performance of a system comprising many components relating to the building systems, fire scenarios and occupant behaviour.

Life-safety outcomes should a fire occur are essentially time-based:

- Fires grow through a series of stages so that fire and fire effluent spread to progressively larger building areas with time
- Escape processes involve a time series of components including detection, warnings and occupant evacuation (or protection in place)

The ultimate evaluation of the performance of the system from a life-safety perspective is to compare the time available for escape with the time required for escape (or tenability time in refuges), since this incorporates the performance of all the separate elements involved in fire development, occupant evacuation and survival [1–4].

For fire engineers this is summarized in the essential principle of performance-based design:

Available Safe Escape Time > (Require Safe Escape Time + an appropriate safety margin)

Where:

ASET = time from ignition to loss of tenability
REST = time from ignition to escape

D.A. Purser (✉)
Hartford Environmental Research, Hatfield, Hertfordshire, UK
e-mail: david-purser@ntlworld.com

A. Cuesta et al. (eds.), *Evacuation Modeling Trends*,
DOI 10.1007/978-3-319-20708-7_2

In this context, tenability limits are exceeded when exposure conditions are such that occupants are no longer able to save themselves. For a defend-in-place strategy it would mean exposure to conditions threatening significant injury or death.

With regard to RSET assessment and simulation it is important to recognise the distinction between "escape" and "evacuation". The evaluation of escape processes and escape time involves everything that happens from the ignition of the fire to the time that all occupants have reached a place of safety. Evacuation processes are a sub-set of escape, consisting of the time from when a warning is presented to occupants to that when they have reached a place of safety. Evacuation time therefore includes pre-travel activities (pre-movement time) including the recognition time (the time taken for occupants to recognise fire cues or warnings and stop their pre-fire behaviours) and Response time (the time taken for occupant to engage in a range of activities related to the fire emergency but before they begin to travel towards the exits). This includes activities such as stopping work processes, seeking information, gathering belongings and warning others.

Since the time periods from ignition before warnings are given (the times from ignition to detection and from detection to warning), and the pre-movement period of evacuation, are most often the greatest part of both total escape time and evacuation time it is very important that these are included in escape or evacuation analysis or simulation. Yet most simulations are limited to the last and shortest process—that of travel to exits and through escape routes.

Although RSET/ASET comparison is usually thought of in the context of performance-based design it is in fact not a new concept but goes back to UK Post War Building Studies and beyond [5, 6].

Around that time committees developing fire safety guidance for buildings considered a number of fire incidents (including the famous Empire Palace Theatre incident, Edinburgh in 1911) [5, 6], and decided that in most cases approximately 2.5 min were available for occupants to escape from when a fire was of a size that it became obvious (a simple estimate of ASET).

They then carried out a performance-based design of escape route capacity, requiring that the aggregate flow capacity (hence width) of available exits should be sufficient to enable the maximum occupant population of an enclosure to evacuate into a protected escape route within 2.5 min [6]. This thereby provided a simple estimate of RSET.

This is still the basis of the prescriptive guidance for exit, escape corridor and stair widths in England and Wales (Approved Document B [7]).

For performance-based design the ASET/RSET methods have become somewhat more sophisticated than the original 1930s application, with more terms in the ASET and RSET expressions and more detailed methods for evaluation of the individual terms and the interactions between them [3, 4].

The original estimates for fire hazard development did not consider methods of and times to detection, fire growth dynamics, or calculation of smoke, heat, effluent toxicity and spread. The original RSET methods did not consider detection, warnings and occupant behaviour times, but only the physical "hydraulic" travel phase of occupant evacuation.

The current Building Regulations for England and Wales (2010) [8], as well as those for Northern Ireland and Scotland are Performance-Based, and the ASET/RSET principle forms the basis of the requirements for means of escape [7]:

"The building shall be designed and constructed so that there are appropriate provisions for the early warning of fire, and appropriate means of escape in case of fire from the building to a place of safety outside the building capable of being safely and effective used at all materials times".

In practice an acceptable design performance can be achieved either by following a set of prescriptive design guidance or by a fire engineering performance-based design.

The same principle is also applied in part to the requirements for fire spread of linings in that they should "have if ignited a rate of heat release or rate of fire growth which is reasonable in the circumstances" and for Internal fire spread (structure) for which "...in the event of fire, its stability will be maintained for a reasonable period."

So the ASET/RSET principle is absolutely fundamental to the required performance of buildings and other structures (such as transport vehicles). Yet almost all aspects of current fire safety design and regulation are based on simple prescriptive requirements for individual features, with limited consideration of interactions between parameters or performance-based evaluation. In the context of this book it is useful to consider how fire safety objectives have been approached and evaluated in the past and what opportunities are presented by current and future performance-based modelling approaches. It is also important to examine the limitations and inadequacies of current methods and how these may be improved by future developments.

Historically most fire safety regulation has been based on changes to individual features in response to particular incidents. Probably the first requirements related to passive fire protection, such as those relating to building construction and separation following the great fire of London in 1666. Passive protection in buildings has been based on fire-resisting materials, initially on an *ad hoc* basis, but increasingly during the last century based upon a testing and certification regime. Furnace tests performance has been used for certification, but originally the relationship between fire-resistance time in a standard test and that in a real fire scenario was only loosely established (2 h fire resistance in a furnace test does not ensure 2 h fire resistance in all potential building fires). In practice the main way to evaluate the success of the standard has been to examine fire statistics over a period since the introduction of the standard, rather than by any performance-based engineering analysis. During the last century there was an increasing progression towards the development of tests designed to provide data for input into fire engineering calculations for structural performance and for the use of "natural" fire curves, more representative of actual fire conditions in buildings. These methods can also be applied to several different structural features so as to enable engineering calculations to predict interactions between different structural elements and the performance of entire structural systems.

So structural fire protection is developing from a set of somewhat *ad hoc* requirements and tests to a more mature and integrated engineering discipline, at least with regard to structural integrity performance. When it comes to parameters affecting occupant escape performance and hazard evaluation such progress has been less obvious. Here there is still a tendency to focus on individual aspects in isolation and to rely on simplistic test and certification criteria rather than obtaining data that can be used for engineering hazard calculations and that can be integrated with data for other elements to enable prediction of the behaviour of building and occupant systems during fires.

An example relates to available and required escape time from dwelling fires. Although the performance-based requirement in the UK is to provide safe means of escape, engineering assessment methods for calculating ASET and RSET are very limited, both with regard to the identification and assessment of individual human behaviour and fire hazard parameters and their integration to assess overall performance. In practice, when problems have arisen, the response had been rather simplistic, with changes to individual features, with the hope of improving performance, but without any performance-based assessment of the extent to which the new features will actually improve safety or interact with other aspects.

Has this approach worked? Well, two examples are changes to furniture flammability requirements and the introduction of smoke alarms, each introduced with the intention of improving fire safety in domestic dwellings. Information on the effects of these changes is revealed in the historical annual United Kingdom fire statistics data for deaths per million population shown in Fig. 2.1. The fourfold increase in deaths resulting from toxic smoke exposure between 1955 and 1970, together with concerns and research on the fire behaviour of upholstered furniture, led to the introduction of the Furniture Regulations in 1988. These introduced controls on the ignitibility of furniture from cigarettes and small flaming ignition sources, which lowered the probability of sustained fires occurring from the application of small fire sources. Fortuitously (but not by design), the changes in covering and filling materials needed to satisfy these test requirements (often involving the use of flame retardants), also resulted in a slower flaming fire growth following exposure to ignition sources large enough to overcome the designed level of ignition resistance. This thereby increased the ASET time available for occupants (especially occupants of domestic dwellings) to escape before conditions became untenable.

Also, from around this time there was an increasing market penetration of domestic smoke detector/alarms, increasing from 9 to 74 % of dwellings between 1987 and 1994. By providing early warning of fire these thereby decreased the RSET time required for escape, when they were effective.

Figure 2.1 shows that the combination of these two measures has fortuitously had a beneficial outcome, so that the fire death rate/million has decreased from 18.01 in 1988 to 7.64 in 2010. However, a more detail examination of the fire deaths and injury statistics reveals a more complicated pattern. The changes to upholstered furniture requirements had no immediate effect, since only a small proportion of householders replace their furniture and beds each year. Despite this, the annual death rate from dwellings fires, especially deaths from smoke exposure, did in fact start to decrease continually from around 1988, but the numbers of fires and the incidence of smoke exposure injuries continued to increase over the next 10 years. After this both the annual numbers of fires and the annual numbers of deaths and injuries continued to decrease up to the present day.

Although other factors may be partially responsible for these improvements in fire deaths and injuries it seems likely that the initial decrease in death rates may have been mainly due to the increasing use of smoke alarms. Occupants received earlier warning of fires and were more likely to survive and escape, but due to the ease of ignition and rapid fire development of the older style furniture, producing large volumes of highly toxic smoke, there was a high probably of building occupants suffering significant smoke exposure while escaping, hence the continued increase in smoke injuries. Once there was significant market penetration of the new-style furniture, the combination of a lowered probability of ignition, coupled with slower fire growth and early warning of fire from the smoke alarms, resulted in a synergistic interaction: increased ASET coupled with decreased RSET, leading to the observed continued decrease in smoke injuries and deaths.

Despite the improvements since the high point of fire deaths in the 1970s and 1980s Fig. 2.1 shows that toxic smoke is still the major cause of fire injuries and deaths, and that we have only succeeded in almost getting back to the rates of injuries and deaths occurring in the 1950s, which represents rather limited progress.

Rather than waiting for the results of these changes to show up in the fire statistics, another way to examine the likely benefits of different interventions could have been to carry out a full ASET-RSET analysis of domestic fires. This has been done for some domestic fire scenarios by carrying out full-scale fire tests in fully furnished test houses coupled with studies of human response behaviour during actual

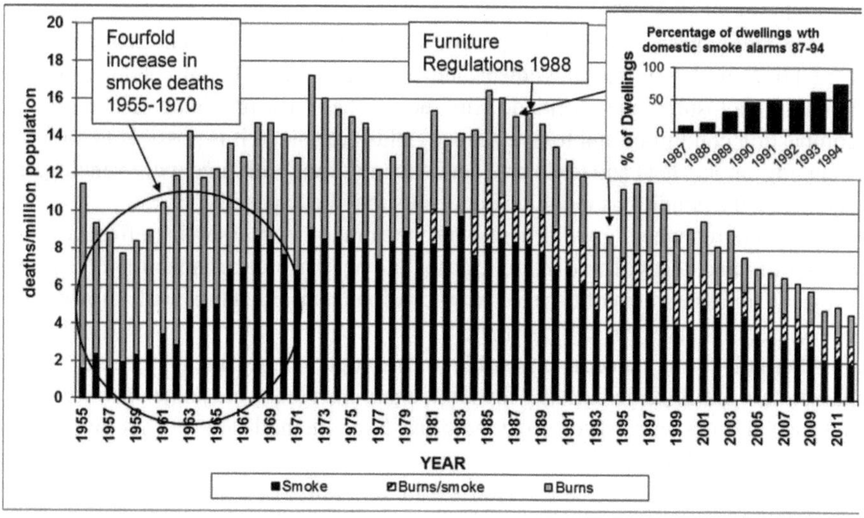

Fig. 2.1 UK Fire deaths per million population [9]

fire incidents. These have revealed the very short time window available for occupants to become aware of alarms and escape before conditions in typical domestic fires can become untenable. However from a design perspective it would be preferable if we could assess such situations by carrying computer simulations of fire development and occupant escape. In order to carry out such a simulation it is necessary to identify each parameter in the ASET and RSET in a fire and escape scenario, determine which parameters have a significant effect on the outcome, quantify them and incorporate them into a simulation. An example of such an analysis is presented later in this chapter for the Rosepark Nursing Home fire.

Attempting such simulations immediately shows that we have some problems, even for a deterministic simulation of a single fire scenario. Firstly, even if we assume flaming ignition of an item has occurred, we have a very limited ability to calculate the subsequent heat release rate and fuel mass loss rate curve in a defined building enclosure. Although attempts are being made to model pyrolysis rates in flaming fires these are still at an early stage and cannot be applied accurately to complex fuel packages such as upholstered furniture, even if the detailed composition and construction are known.

So currently it is usual to rely on standard heat release rate curves or experimental data from large-scale calorimeter tests, the data from which are limited only to the item tested and for which the test boundary conditions are likely to differ considerably from those in a real enclosure fire.

Another vital parameter in fire hazard development is the changing yields of smoke and toxic products from burning fuels under the different changing combustion conditions occurring in full-scale building fire scenarios. Ability to model these in fire simulations is currently very limited. Experimental studies have shown that yields of key toxic gases including CO and HCN vary considerably depending upon fuel composition and combustion conditions, and this topic is discussed with examples later in the chapter.

One area where considerable progress has been made in both zone and CFD simulation models is the movement and mixing of fire effluents plumes through building enclosures once formed. This means that once problems or fire growth, combustion conditions and toxic product yields are solved it is possible to simulate changing smoke, heat and toxic product concentrations with time within enclosures, and hence the conditions to which occupants may be exposed.

Another area where good progress has been made is in assessing effects of concentrations and doses of smoke, toxic gases and heat on escape capability of exposed subjects, for assessment of time to incapacitation in specified fires with known time-concentration curves for key parameters.

Alternatively, as stated, it is possible to carry out and measure all the relevant fire parameters in full-scale building fires, but again the results are somewhat limited to the specific fuels, enclosures and conditions. Such methods are very useful for validation of modelling predictions.

Even when these issues are solved we are left with the enormous problem of the wide varieties of fuels, building contents, structures and configurations occurring in practice and the probabilistic aspects of fires involving them. Our ability to make realistic assessments of fire hazard scenarios and their probability of occurrence could be considerably improved by more assiduous collection of some fairly basic data. For example, in different building occupancy categories, what proportion of interior doors and windows to the exterior are open or closed at different times, and at what times during actual fire incidents—a parameter having an enormous effect on tenability time during fires? What are the average and range of compositions of television sets or washing machines in terms of the chemical elements important in toxic fire hazard, or for whole contents of different rooms or vehicle interiors? It is noticeable in relation to all these parameters, that efforts to collect data, develop calculation and simulation methods all tend to be heavily directed at some parameters, while others are largely ignored.

Similar issues arise when RSET parameters are considered. Quite a lot of effort has been directed at measuring and modelling the response of smoke and heat detectors during compartment fires. However very little research or data are available for the probability and timing of occupant recognition and response to warnings. As the Dusseldorf Airport example described later illustrates, this is especially a problem when occupant warnings depend on the response of security personnel to pre-alarms or warning. In many such incidents, security and management personnel were aware of fire detection for many minutes before evacuation warnings were delivered to other occupants, sometimes resulting in multiple deaths.

In contrast to this, much effort has been directed over the years to the physical design of means of escape and the calculation and simulation of the travel phase of escape and evacuation. Although it is important to develop a good basic travel and flow simulation component as the basis for any evacuation model, it seems to me that we are approaching diminishing returns in this area, in that almost all research and development in human behaviour is devoted to small improvements in simulating this parameter, while other equally important terms on the RSET equation are almost ignored. A prime example is the World Trade incident, for which evacuation travel capacity was not the main limitation on escape and survival, but the very long delays in warning occupants to evacuate and in their response to the developing fire cues, which resulted in delays of around and hour before many occupants decided to start evacuation travel.

If we consider for example a large building, or tunnel incident in which warning and pre-movement can be estimated as 20 ± 15 min and physical evacuation travel time as 5 ± 0.1 min, how much are we going to improve our total escape time simulation accuracy by improving our travel time calculation accuracy to 5 ± 0.01 min?

Of course it is much easier to simulate physical movement than human response, and it is much easier to conduct and measure experimental evacuation travel drills than to set up and perform unannounced escape and evacuation experiments to collect this vital data.

So what are the challenges for future development of fire hazard, escape and evacuation simulations?

- Examination of the entire escape and hazard development process involving all interactions between the occupants, built systems and fire scenarios.

- Identification of those parameters having the greatest effect on safety outcomes.
- Design studies and experiments to collect the necessary data on the key ASET and RSET parameters and to design functions and simulation methods to replicate their behaviours.
- Design simulation methods that replicate the main features of each key parameter and integrate them into a functioning system enabling prediction of outcomes for specified deterministic scenarios.
- Collect data on the range of variability of each key parameter to facilitate probabilistic analysis of simulation outcomes.

Specific issues relating to escape simulations that require research and improvement are:

- Measurement of response behaviours and decision times in reporting chains from persons first discovering a fire, and through management, leading to general evacuation warnings to occupants. Development of simulation functions for these behaviours and incorporation into simulation models.
- Measurement of recognition and response pre-movement behaviours of people in different kinds of occupancies in relation to different warning systems and fire safety management practices. Development of simulation functions and probability distributions for these behaviours and incorporation into simulation models.
- Measurement from experiments and collection of data from fire incidents on behavioural responses and decision making in response to seeing or being enveloped in smoke. In particular, under what conditions are occupants likely to take refuge rather than escape (for example remaining in a bedroom when the landing is smoke-filled, or in their vehicle during a tunnel fire)? Under what conditions (at what smoke densities) and in what scenarios are they likely to decide to travel through smoke? What are probability distributions for such decisions, for example as a function of smoke density or proximity to the fire?
- Vital aspects such as these are usually ignored in simulations, although some useful experimental research has been carried out on the effects of smoke density on travel speed once occupants decide to evacuate, especially in tunnels. Although these data can be incorporated into evacuation simulations, they may be of limited relevance if most occupants would in fact remain in their vehicles as occurred during the Mont Blanc tunnel fire. Criteria are therefore needed for this decision process for incorporation into simulation models.
- With regard to travel and flow simulations for occupant populations through escape routes, existing physical simulations models are very effective, but further development and validation are needed with respect to merging behaviour and flows. This has been shown to be partly a physical problem and partly dependent on aspects of social behavioural interactions. Another aspect of travel which is behavioural rather than physical, but which can have a large effect on evacuation flows is exit choice and wayfinding behaviour. In addition to collecting data on this aspect there is need for the development of a functional basis for application in simulations.

So as I see it, the twentieth century saw the birth of performance-based analysis and the development of engineering calculation methods for individual fire parameters. The challenge for the first part of the twenty-first century is

- To recognise and improve understanding and methods for calculations of important parameters previously neglected in simulations.
- Develop improved and more comprehensive simulation methods to integrate the effects of different parameters interacting in systems.
- Collect and apply sufficient data on variability of input data to improve the capability of probabilistic simulations.

The remainder of this chapter discusses the benefits and possible alternatives to ASET/RSET comparisons and simulations and presents three examples of the investigation and application of ASET/RSET components and interactions. The text is the basis for the 2013 Rasbash Medal Lecture presented by the author at the Institution of Fire Engineers 2013 Conference, Stratford-upon-Avon 10–11 July 2013.

2.2 Is There an Alternative to ASET/RSET Comparisons?

So given that ASET/RSET comparisons form the basis of the prescriptive guidance and the performance–based engineering standards, I was somewhat surprised and concerned to come across a 2010 paper published in Fire and Materials entitled " RSET/ASET, a flawed concept for fire safety assessment". I have also found that some international colleagues in the fire safety engineering standards arena seem to have a limited understanding or appreciation of the importance of this principle.

This gave me pause for thought—have the codes and engineering standards been wrong all this time and what possible alternative might there be to ASET/RSET analysis for the fire safety performance of buildings?

The paper cited above recounts a number of fire incidents in dwellings in which occupants have failed to escape in safety or have attempted to re-enter burning buildings, and more importantly, design cases for which the ASET/RSET methods have been incorrectly applied. The paper concludes by recommending ASET/RSET calculations with use of appropriate safety margins—which is of course the original basic concept. So perhaps the title is somewhat misleading in that it is not so much the basic concept that is flawed as its application, and the paper does a useful job in drawing attention to such misuse.

Considering the whole issue I feel that there are alternative methods for improving life-safety other than the application of ASET/RSET comparisons. As mentioned with respect to the Furniture Regulations one method is to introduce measures that target specific issues and reduce the probability of occurrence of specific failure modes. Thus the implementation of the furniture regulations reduced the probability of occurrence of furniture fires in general. The introduction of domestic smoke alarms reduces the probability that domestic fires will remain undetected until conditions become untenable. However, risk overall is the product of probability of occurrence and severity of the ensuing hazards. Also, when considering the overall performance of built systems, there is a general probability of fires starting in a variety of components for a variety of reasons. For any situation in which a fire does occur it then becomes important to consider the time-based performances of the entire system to evaluate fire safety, hazards and risks to occupants.

2.3 Benefits of ASET/RSET Comparisons

I am convinced that, carried out correctly, time-based ASET/RSET comparison provides a very powerful method for evaluating and improving fire-safety for the following reasons:

- It evaluates the overall performance of the whole system
- It forces the designer and regulator to identify and consider the performance of each individual component of fire hazard and means of escape—and justify any assumptions
- It enables the influence of each parameter on the performance of the whole system to be evaluated. Areas of weakness can then be addressed
- It enables identification of parameters having a large influence on system performance and where redundancies or overlaps occur. (For example time to detection is an additive term in RSET and therefore very important. Pre-movement and travel times, although additive for individual occupants, overlap for evacuating populations—Fig. 2.2)
- It enables trade-offs between different components to achieve desired overall performance
- It enables development and evaluation of defence-in-depth, to determine effects on overall safety outcomes when one or more components fail.

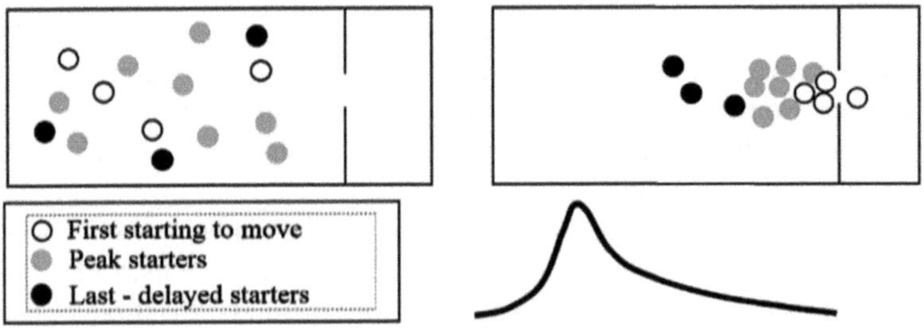

Fig. 2.2 Pre-movement time distributions for occupant populations follow log-normal distributions but for a crowded enclosure the evacuation time depends upon the pre-movement time of the first few occupants (who form the queue at the exit) plus the travel flow time for the population. Late starters simply join the end of the queue at the exits [12, 13]

- It highlights important parameters for which we have insufficient knowledge—providing a focus for meaningful research (e.g. merging flows at storey exits), or for the application of safety margins.
- For fire incident investigation the construction of a time-line for fire development and occupant behaviour is vital to understand what happened—the majority of cases involve ASET/RSET failures in terms of fire and escape design and management

Comparing the interactions between different parameters as in Fig. 2.2 enables optimisation of means of escape and effects of trade-offs to be calculated.

2.4 Three Examples of the Investigation and Application of ASET/RSET Components and Interactions

In order to illustrate the benefits of investigation and application of ASET/RSET components and system interactions I have chosen three different examples from different topics I have worked on:

1. RSET topic: "pre-warning" delays in evacuations of large buildings with two-stage alarm systems resulting in life loss in major fire incidents
2. ASET topic 1: Using comparisons of forward-calculated Fractional Effective Dose and carboxyhaemoglobin concentrations with measurements in incident victims to validate fire test or computer modelling incident reconstructions
3. ASET topic 2: combustion chemistry: Non-conservative constant values are often used in fire dynamics models for heats of combustion and toxic gas yields. The ISO/TS19700 tube furnace—developed at BRE—can be used to measure these variables as a function of fuel/air equivalence ratio [14–16].

2.4.1 RSET Topic: Pre-warning Delays in Major Fire Incidents

The basic RSET expression for a simple building is as follows:

$$RSET\left(\Delta tesc\right) = \Delta tdet + \Delta ta + \Delta tpre + \Delta ttrav$$

Where

RSET (Δtesc) = escape time (time from ignition for all occupants to reach a place of safety)
Δtdet = time from ignition to detection
Δta = time from detection to alarm
Δtpre = pre-movement time
Δttrav = travel time

For many smaller buildings a simple (Level 1) automatic detection and alarm system is used whereby detection usually results within a few milliseconds in a general evacuation alarm or warning to all occupants. In such a simple system alarm time is effectively zero. After a pre-movement period during which occupants recognise and respond in various ways to the alarm they then travel towards the exits and evacuate [2, 3].

Larger building often have two-stage alarms (Level 2 systems). Detection activates a pre-alarm to security after which a general alarm is sounded to all affected occupants. The time from ignition to the general alarm is the simple sum of the detection and alarm times. But in many larger and more complex buildings, when the initial detection of the fire (either automatic or by a person) provides a pre-alarm to security staff, they then have to decide if and when to activate a general evacuation alarm to occupants of affected enclosures (or the entire building). Sometimes there is an automatic protocol for a general alarm (for example after a fixed delay unless cancelled, or if more than one detector is activated by the fire), but in some serious incidents either there has not been such a back-up or it has been cancelled. When staff receive the pre-warning they enter their own pre-movement sub-routine, taking time to recognise that an alarm has been received and then responding in some way. In almost all cases the first response is to send someone to investigate. Further time is then spent for investigators to travel to the fire site, appraise the situation and report back. Investigation and reporting may take a significant time as the fire grows before it may be considered sufficiently serious to recommend evacuation. Information on the developing scenario and instructions may pass up and down a management chain. Eventually, if the fire becomes more serious, a decision may be made to active a general evacuation warning to affected occupants, but in some incidents this has occurred too late, so that members of the occupant population suffer a serious exposure while attempting to escape.

In such a situation the RSET expression might be represented as follows:

$$RSET\left(\Delta tesc\right) = \begin{pmatrix} \Delta tdet + \Delta tpre \\ \uparrow \longleftarrow \downarrow \end{pmatrix} \longrightarrow \Delta ta + \Delta tpre + \Delta ttrav$$

Where

Δtpre represents the pre-warning time

In practice in such incidents it is possible to recognise a kind of sub-routine consisting of a sequence of different "detection" stages or events leading to a sequence of different "pre-warning" events before the final pre-warning leads to an alarm [17, 18].

Issues such as these have made an important contribution to the consequences of a number of serious fire incidents including:

- Dusseldorf Airport
- Manchester Woolworths,
- Summerland Holiday Centre
- Nagasikiya store fire, Amagasaki City, Japan

- Bradford Stadium,
- Rosepark Nursing Home
- Dupont Plaza and MGM Grand Hotel Fires
- Daegu Korea and London Kings Cross Underground Stations
- Channel Tunnel and Mont Blanc tunnel
- World Trade Centers
- Stardust Disco
- Lakanal House

The Dusseldorf Airport Germany 11th April 1996 [19] provides an example of one such incident: (Table 2.1). This fire started as a result of "hot work" (welding) on an access roadway above one of the terminals, with the fire penetrating down into a void and then through the terminal ceiling of the Arrivals hall near a flower shop (Fig. 2.3). The fire was detected as a smell of smoke and visible sparks by a taxi driver waiting on the taxi rank. He reported the fire to airport control by telephone. Two minutes later, two airport firefighters were sent to investigate. They called an electrician to investigate an electrical burning smell, but the fire and smoke gradually worsened so that after 12 min a full attendance of the airport fire service was requested. After 16 min a firefighter saw a rapidly spreading glowing area over the suspended ceiling and instructed 20 persons to leave the nearby cafeteria, but no general alarm is given. After approximately 27 min a flashover occurred and the automatic general evacuation alarm was triggered, as well as a fire shutter, which descended too slowly to prevent fire spread to an adjacent hall. The evacuation alarm consisted of a taped message instructing occupants to evacuate towards the fire area. There were 16 deaths including eight in the Air France VIP lounge, five in a lift which descended from the car park and opened on the arrivals hall fire, one person in a toilet and two in the main arrivals hall.

So for this incident:

ASET

Arrivals hall = ~16 min ceiling collapse 30 min flashover
Air France lounge = 23 min smoke blocks escape routes
Departures hall = 31 min smoke spreading at walking speed

Table 2.1 Timeline for Dusseldorf Airport incident

15:31	0	Taxi driver reports smell of smoke and sparks on taxi rank roadway
15:33	2	Two airport fire fighters send to investigate. They called electrician
15:40	9	Electrician attempting to identify source of smoke
15:43	12	Smoke worse: full airport fire service requested
15:45	14	Occupants notice strong stench of burning rubber some leave
15:47	16	Firefighters sees rapidly increasing area of glowing to 20 m^2 of suspended ceiling, shouts to 20 people in Cafeteria to leave (no alarm sounds) then ceiling collapses
15:50	19	Occupants see large cloud of black smoke above arrivals level parking garage
16:06	27	Flashover in hall activates fire alarm and shutters. Automatic voice alarm message directs occupants to evacuate towards the fire

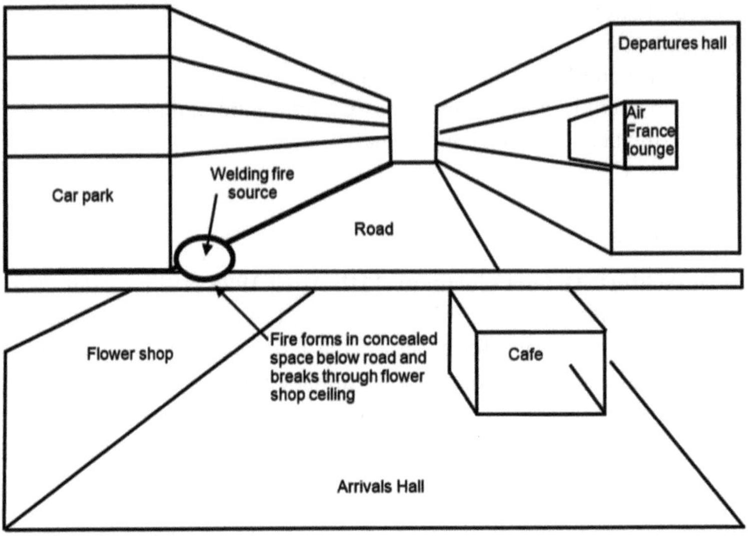

Fig. 2.3 Dusseldorf Airport

$$RSET\left(\Delta tesc\right) = \left(\Delta tdet + \Delta tpre\right) \rightarrow \Delta ta + \Delta tpre + \Delta ttrav$$

Δtdet + Δta Taxi driver 0 min, before or soon after flaming ignition?
Δtdet Airport Fire fighters 2–16 min (long pre-warning period)
Δtpre Airport Fire fighters 26 min Tell airport security to clear terminal
Δta Voice alarm message 35 min (5 min after flashover and directs towards fire)

The general findings from this and other incidents are that both "detection" and "pre-warnings" are not simple processes, even when sophisticated detection systems are present. For the transport case described, and others, the affected enclosures contained many occupants (travellers) or involved staff, so that cues to all fires were observed by some persons soon after ignition, when the fires were quite small [17, 18]. In two cases (King's Cross and Dusseldorf) the first person to discover the fire very quickly reported it to the local authority, but in other cases (Channel tunnel, Daegu, Mont Blanc) they either failed to communicate quickly or as in the case of several incidents, were slow to interpret the fire or smoke cues as representing a potentially serious fire and issue timely evacuation warnings, so that there was an extended "detection" period interspersed with an extended "pre-warning" period. A complication with Mont Blanc was that the "opacimeters" providing cues to the fire in the control room were not intended as a fire detection system.

Designer issues: For enclosures such as stations or train carriages, toilets etc., occupancy cannot be guaranteed, so automatic detection is necessary, but for normally occupied enclosures obvious manual means of reporting a fire and raising an alarm should be provided. For large enclosures with high ceilings early detection by occupants is even more likely. Careful consideration needs to be given to the likely series of cues that occupants and staff will become aware of, and how they will and should be trained to interpret and act on the information received, especially what constitutes "detection" of a serious fire.

With regard to the provision of general evacuation warnings it is very important at the design stage (and in simulations) to consider the nature and timing of information regarding a fire that will become evident to security and control staff.

It is then necessary to consider the sequence of investigation activities the staff will be expected to engage in and the sequence of communication and decision making up and down the management chain.

- How serious does an incident need to be before security staff can activate a general evacuation warning?
- What autonomy do passengers or staff at the fire scene, or receiving information directly from it, have to initiate a general evacuation?
- Given that fires in transport incidents may develop from minor to severe fires within a few minutes, how long should be allowed in the design for the detection and warning processes to take place? How can effective firefighting resources be activated to arrive at a fire scene within a few minutes of the first detection?
- For transport situations how is fire information provided to station and train or tunnel operation management to enable facilities to be closed so that no further trains or vehicles enter the danger area?

2.4.1.1 ASET TOPIC 1: Validation of Rosepark Full-Scale Reconstruction Fire Using Fractional Effective Dose Modelling and Forensic Incident Analysis

The Rosepark nursing home fire occurred at 04:28 h on 31st January 2004 and resulted in 14 deaths [20]. As part of the investigation on behalf of the Scottish Office and the Procurator Fiscal a full-scale test reconstruction of the fire was carried out by BRE [21]. The author participated in this work and also investigated the fire time-line and effects on decedents and exposed survivors using documentation related to the incident.

As part of the BRE work time-concentration curves for smoke, toxic gases and temperature from the fire test data were used to calculate (simulate) the uptake by exposed occupants in different locations and the predicted toxic effects and burns. This analysis took the form of Fractional Effective Dose (FED) calculations, whereby the received dose of each toxic gas (or heat) during the fire is expressed as a fraction of the dose predicted to cause incapacitation. The FED of an occupant increases with time during the fire as they are exposed to a toxic gas until a point is reached when the FED exceeds 1, at which time incapacitation is predicted. As part of the analysis the uptake of carbon monoxide was also calculated in terms of calculated percentage carboxyhaemoglobin (%COHb) in the blood, and also the extent of predicted pain or burns from heat exposure. By this means forward-calculated

%COHb concentrations were predicted for each occupant at the time of incapacitation or rescue (for survivors) and the time at which a lethal concentration was achieved (for decedents). This information was used to estimate hazard development during the incident, but in practice the findings were only as good as the extent to which conditions in the reconstruction fire were similar to those in the actual incident, and the extent to which the FED calculation input data and models were representative of the uptake of toxic gases and effects on the actual occupants.

In practice considerable care had been take to replicate the important structural features, fire fuel sources and boundary conditions of the actual incident, and the pattern of fire damage in the experimental rig after the test was very similar to that found during investigation at the incident site.

Another source of information used to validate the test results was the measured %COHb in the blood of incident decedents and survivors. For the 10 persons dying at the fire scene during the incident the %COHb levels represented those at the time of death. For the 8 persons rescued alive from the fire scene the %COHb levels were those recorded from blood samples taken soon after arrival at hospital. After rescue these subjects were treated with oxygen, which gradually washes CO from the blood, but if the time of rescue, subsequent treatment and time of blood sampling are recorded it is possible to back-calculate from the %COHb concentration measured in hospital to that in the blood at the time of rescue.

The extent to which the forward calculated %COHb from the fire test data agrees with the %COHb at death and the back-calculated %COHb in the blood of survivors (and the extent of predicted burns compares with actual burns) therefore provides an indication of the extent to which the fire test conditions and FED calculations are representative of the actual fire incident conditions and effects.

Figure 2.4 shows a plan of the upper (ground) floor of the nursing home which was built on a sloping site. The fire occurred in an open cupboard marked with a star on the plan. A brief violent fire occurred which filled the dog-leg corridor and open bedrooms off it with a dense toxic effluent. Figure 2.5 shows the short high temperature fire in the cupboard, and there was a similar temperature profile in the corridor. However as Fig. 2.6 shows the temperature in the open rooms did not exceed 150 °C at bed height and then only for a brief period. The fire self-extinguished after approximately 6 min as the oxygen in the enclosed system became depleted.

Figure 2.7 shows the time-concentration curves for the smoke and toxic gases in the corridor, which were similar to those in the open rooms off the fire corridor. The FED analysis and calculated %COHb concentrations for an occupant of the fire corridor are shown in Fig. 2.8. The analysis shows that a corridor occupant would be seriously affected by dense irritant smoke from 4 min (as the FIC smoke curve crosses 1 on the Y-axis). Incapacitation (loss of consciousness from the effects of asphyxiant gases) is predicted at 5.5 min (6.5 min in the open bedrooms at bed height). Pain from heat exposure is predicted after 6 min in the corridor, but neither pain nor burns in the open bedrooms. The %COHb increases rapidly after 6 min and in the occupied open bedrooms is calculated to exceed lethal threshold levels (50 % COHb) after 7.9 min.

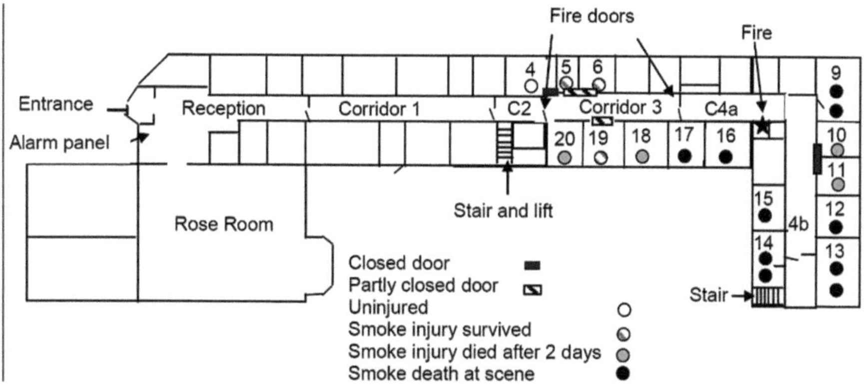

Fig. 2.4 Rosepark Nursing Home ground floor showing locations of exposed residents

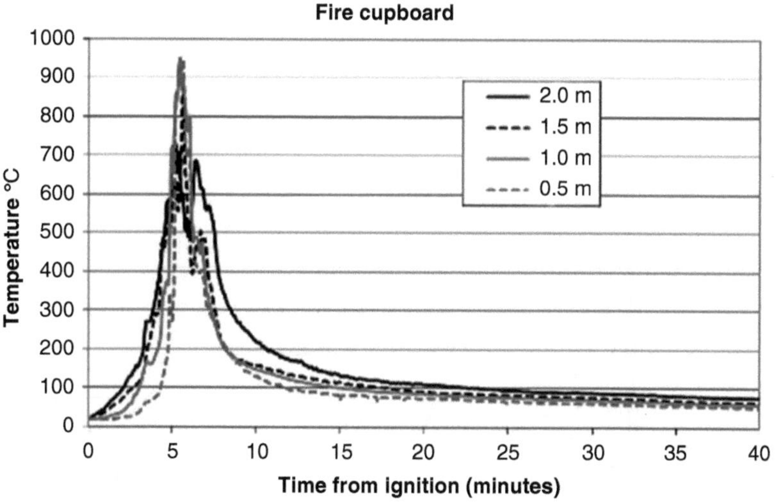

Fig. 2.5 Temperature profile in fire cupboard

From this analysis it was predicted that all occupants of the open bedrooms of the fire corridor would have been incapacitated after 6.5 min and dead by around 8 min. They were predicted to have no burns (at least before death) and that the %COHb concentrations in the bodies would be at lethal levels, exceeding approximately 50 % COHb but with final concentrations depending upon when each individual stopped breathing (ranging from approximately 50–85 % COHb).

In practice all the open room occupants were found dead at the fire scene when the fire service personnel entered the rooms 30–60 min after ignition. They were not burned (apart from some minor superficial, probably post-mortem, burns in a room close to the fire area) and all had very high %COHb concentrations ranging from 48–82 % COHb (mean 63 %). These results were therefore consistent with the

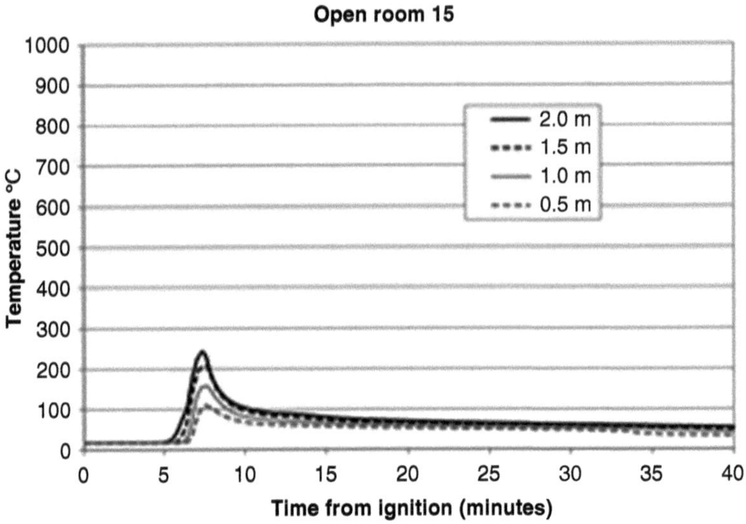

Fig. 2.6 Temperature profile in open bedroom

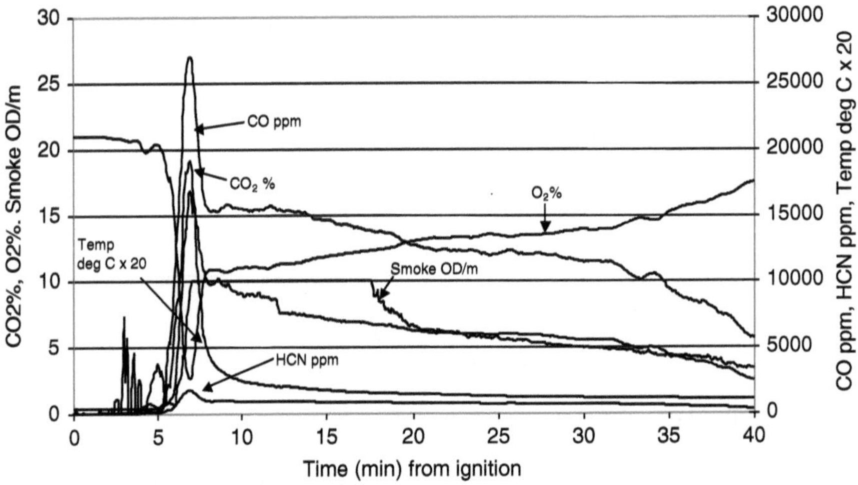

Fig. 2.7 Smoke, toxic gases and temperature profile in the fire corridor during the BRE full-scale recreation of the Rosepark Nursing Home fire. Conditions in the open bedrooms off this corridor were similar except that temperature at bed height was considerably lower

findings predicted from the fire test and FED modelling. However, because the exact times of death could not be established from the forensic data it was not possible to fully validate the CO concentrations and uptake in the test against the actual fire incident conditions.

As indicated in Fig. 2.4, two occupants were in closed rooms (10 and 11) off the fire corridor. The door to Room 10 remained intact, while that of Room 11 was

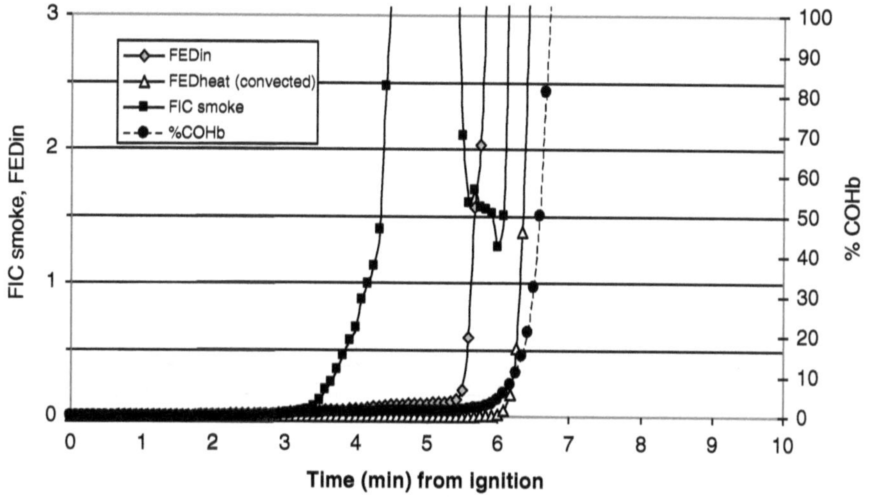

Fig. 2.8 Fractional effective dose hazard analysis for the corridor fire profile in Fig. 2.7

partly burned through during the fire. Since the time of rescue was known it was possible to forward calculate the predicted %COHb at rescue for these occupants from the CO concentration curve measured in a closed room during the fire test, and compare it with the back-calculated values from the blood levels measured after arrival at hospital. Adjacent to the fire corridor was another corridor protected by a fire door which closed when the fire was detected. Unfortunately, the fire cupboard contained some aerosol cans. Each time one of these exploded during the fire the resultant pressure pulse blew open the fire door enabling some toxic smoke to enter the corridor beyond and the open rooms off it. Five occupants were rescued from the rooms off this corridor, with varying degrees of exposure and measured blood %COHb concentrations depending upon the extent to which their bedroom doors were fully or partially opened, or closed during the fire. Similar comparisons could therefore be made between the %COHb concentrations calculated for these individuals from the test data and their concentrations at rescue, back-calculated from the hospital blood data.

Figure 2.9 shows the smoke, toxic gases and temperature profiles measured in a closed room off the fire corridor during the BRE test. Figure 2.10 shows the forward calculated blood %COHb concentration for the occupant of Room 10 up to her rescue after 70 min. The figure shows two uptake curve estimations. These represent the range of uncertainty in the uptake calculation depending upon the respiration of the subject. The lower curve is for a subject at rest breathing 6 L of air each minute (for example a subject resting or sleeping in bed). The upper curve is for a more active subject awake and sitting of standing during the exposure period. The two estimates range from 42 to 56 % COHb at the time of rescue. In practice it is known that this subject got up from her bed and moved around briefly and then sat on a chair, awaiting rescue, before becoming unconscious. Since she was rescued

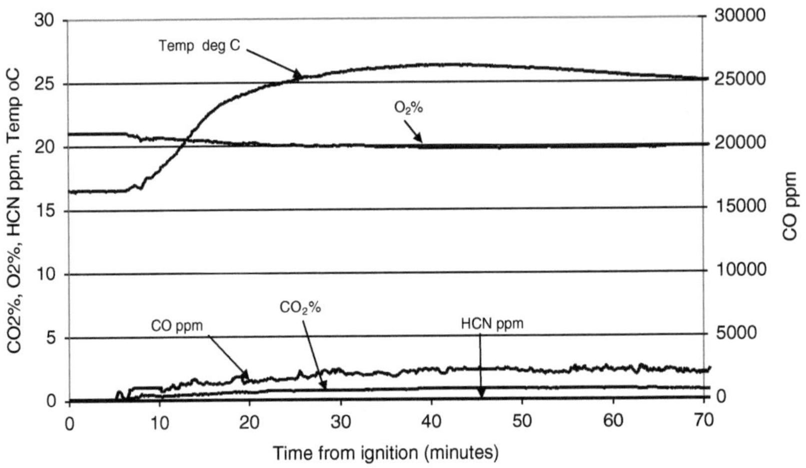

Fig. 2.9 Smoke, toxic gases and temperature profile for a closed room off the fire corridor during the BRE full-scale recreation of the Rosepark Nursing Home fire

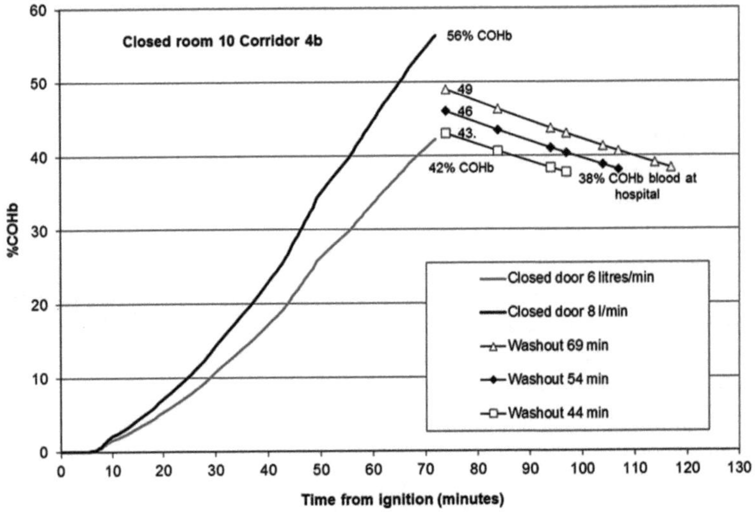

Fig. 2.10 Forward calculated %COHb level for the occupant of closed room 10 from the CO and CO2 gas profiles in Fig. 2.9, assuming sleeping and standing VE values of 6 and 8 L/min compared with %COHb levels back-calculated to the time of rescue using three estimates of the time when the blood sample was taken after arrival at hospital

alive and recovered consciousness in the ambulance it is considered unlikely that she could have had a blood level exceeding 50 % COHb at rescue, so the best estimate would be around 45 % COHb. When a sample was taken at hospital she had a blood level of 38 % COHb. Unfortunately the exact time the blood sample was taken was not recorded, but from other cases it is considered that a sample would have been taken soon after arrival and three estimates of 44, 54, and 69 min after rescue have been made. From the washout back calculations these give a %COHb range at the time of rescue between 44 and 49 % COHb. The best estimate based upon 54 min is 46 % COHb (excluding the 49 % COHb estimate which would most likely have been lethal at the fire scene).

Although both the forward and back-calculated values have some ranges of uncertainty it is considered that the values predicted from the test are very close to those calculated from the blood data and therefore provide a considerable degree of confidence that:

- The conditions in the BRE test fire (the fire size and gas concentrations in the fire corridor) were similar to those in the actual incident
- The leakiness of the closed bedroom doors in the BRE test rig were similar to the leakiness of the actual Rosepark closed bedroom doors, so that the time-concentration curve for CO in the closed test room was similar to that in the actual Room 10
- The FED and %COHb uptake calculations, and burns model, are validated to the extent that the calculated and measured %COHb concentrations were close to those calculated from the measured gas concentrations and the effects on the exposed subject were as predicted (loss of consciousness but survivable at the fire scene with no burns or reported heat pain)

For the occupant of Room 11 forward calculations are complicated by the fact that her room door burned through partially during the fire, which would have allowed the CO concentration in the room to increase considerably to a level close to that in the corridor. Also, since she was rescued at a time when the corridor CO concentrations were still high, she would have suffered some exposure while being carried out along the corridor. When her blood level was forward calculated to the time of rescue plus an additional value for the brief, high concentration, exposure in the fire corridor (but assuming her room door remained intact before rescue) then the calculated value of 34–40 % COHb was somewhat lower than the back-calculated blood value of 43–57 % COHb (Table 2.2). However, when allowance was made for partial door burn through, with increased CO leakage from 34 min to the time of rescue at 41 min, the calculated value increased to around 45 % COHb which is consistent the value using the measured blood data. Table 2.2 summarizes the %COHb comparisons for eight subjects in different locations. For each subject the fire test CO data and times exposed have been used to forward calculate the %COHb at the fire scene. The times on oxygen and blood data %COHb at Hospital (shown in the last column) have been used to back-calculate the %COHb range at the fire scene (shown in the penultimate column).

For the occupants of rooms off the corridor beyond the compromised fire door the forward calculations are complicated in that three subjects had room doors party open or open for part of the time, although two had open doors. For all these subjects the forward-calculated values from the BRE test are somewhat lower than the back-calculated values from their measured blood data. It was concluded that the extent of smoke and CO contamination of this corridor and the rooms off it was somewhat greater in the actual incident than in the BRE test. During the BRE test an extract duct system was omitted which had been present at Rosepark, with

Table 2.2 Occupants alive and rescued at the fire scene: outcomes and comparison between actual and calculated %COHb

Subject and location	Room	Time exposed in room (min)	Time on oxygen (min)	COHb at scene		
				From fire test data	From blood data	
					At scene	Hospital
Closed rooms off fire corridor (corridor 4)						
Door closed, unconscious, recovered in ambulance, pneumonia death	10	72	23–33	42–56	43–49	38
Door partly burned, coma, cardiac arrest, no recovery, pneumonia	11	41	44–69	34–40	43–57	25.8
Open, ajar and closed rooms off corridor 3 beyond fire door						
Door open, coma, resp arrest, no recovery, pneumonia	18	38	51–66	22–29	44–53	24.7
Door open, conscious, pneumonia death	20	27	62–73	20–26	42–55	29.6
Door ajar, conscious, survived	5	32	67–82	~12 19–24	29–32	19.6
Door ajar then closed, conscious, survived	6	36	55–70	~12 22–27	35–38	25.5
Door ajar, comatose, recovered, survived	19	32	67–82	~12 18–24	38–41	24.8
Door closed, uninjured	4	29		~12		

openings to the cupboard and this corridor. Subsequent BRE tests established that a significant quantity of smoke would have been extracted from the fire cupboard into this ducting system, and that some of it would have been released into the corridor beyond the fire door (as reported by witnesses at the time of the fire). This would have added to the total CO concentration in this corridor and rooms off it, which may partly explain why the exposure during the incident appears to have been greater than predicted from the original BRE tests.

Taking these results as a whole, it is considered that the application of the FED and %COHb ASET modelling calculations added considerably to the understanding of the fire conditions at Rosepark, and taken together with the fire test results, incident investigation and toxicology data from exposed subjects, provided a set of evidence validated using the different methods applied. The findings also provide validation of the FED calculation methods to predict effects on subjects exposed in incidents such as this.

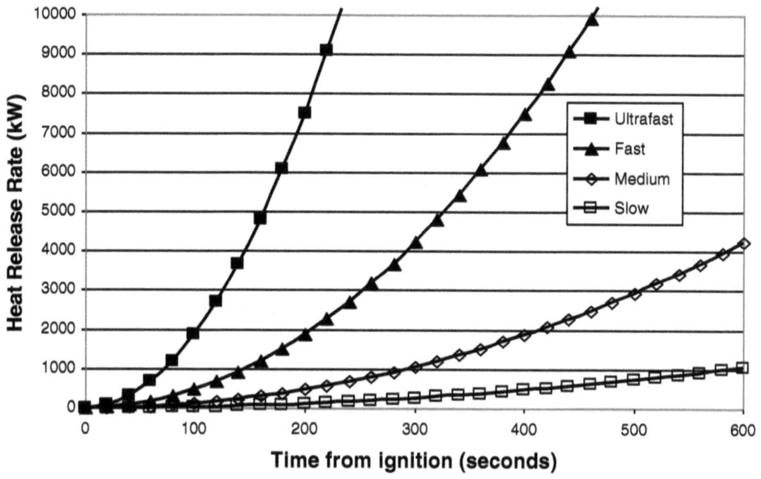

Fig. 2.11 t^2 Heat release rate curves

2.4.1.2 ASET Topic 2: Combustion Chemistry—Measuring Effective Heats of Combustion and Toxic Product Yields as a Function of Combustion Conditions for Input to Fire Dynamics Models

The developing toxic hazard in a compartment fire depends upon the time-concentration curves for the toxic products. This in turn depends on:

- Fire growth curve (mass loss rate of fuel [kg/s])
- Yields of toxic products under a range of combustion conditions (e.g. kg CO/kg material burned)

In fire engineering calculations and computer models the size of the growing fire is often expressed in terms of a t^2 HRR (Heat Release Rate) curve (Fig. 2.11): The mass pyrolysis rate of the fuel is then given by HRR (MW) x heat of combustion (MJ/kg) = kg/s fuel mass pyrolysed. The values used for the heat of combustion are often the constant heats of complete combustion [22, 23]. In practice the actual heats of combustion from burning fuels are likely to be close to these values during the early stages of well-ventilated fires. However, during enclosed or partly enclosed compartment fires typical of fires in buildings, combustion efficiency decreases considerably as the ventilation becomes limited and the combustion fuel-rich. The effective heat of combustion then decreases. This means that to provide a given HRR value inside the fire compartment a greater mass of fuel is pyrolysed.

Similarly, the values for the yields of toxic products commonly used as input to these calculations are often constant values measured during well-ventilated combustion. As the combustion becomes less efficient, the yields of the toxic products of incomplete combustion including smoke particulates, irritant organic species, CO and HCN are considerably increased.

The results of these two common oversimplifications are non-conservative. For example halving the effective heat of combustion results in double the mass production rate of the fuel and toxic products, and when this is accompanied by a tenfold increase in CO yield per kg of fuel mass pyrolysed the result is a 20× increase in the mass production rate of CO.

The ISO/TS19700 tube furnace was developed at BRE to measure heats of combustion and toxic product yields from fuel materials and products over a range of fire conditions. These data can then be used for input to fire simulations to enable calculations of time-concentration curves of fire effluent species [14–16].

The method has been validated for a number of fuels against the yields obtained in large-scale and full-scale compartment fires [24–27].

For flaming combustion, heats of combustion and effluent species yields are measured as a function of the fuel/air equivalence ratio (ϕ). Under well-ventilated combustion conditions ($\phi < 1$), toxic product yields for non-flame retarded materials are low, but increase considerably under fuel-rich conditions ($\phi > 1$).

$$\phi = \frac{Actual\ \ Fuel\,/\,Air\ \ Ratio}{Stoichiometric\ \ Fuel\,/\,Air\ \ Ratio}$$

The effects of other variables, including temperature, oxygen concentration and non-flaming and flaming decomposition are also measured.

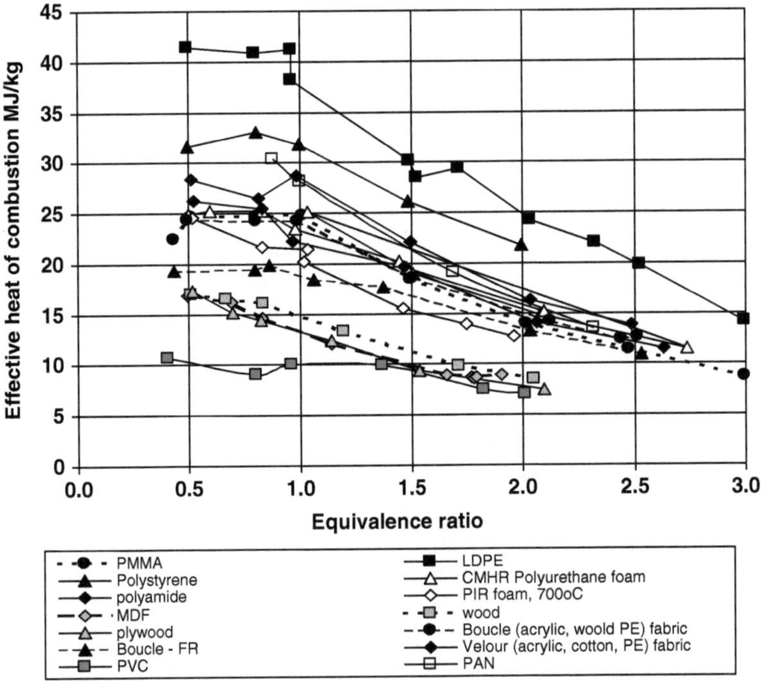

Fig. 2.12 Relationship between equivalence ratio and effective heats of combustion measured using the ISO/TS19700 tube furnace [24, 25]

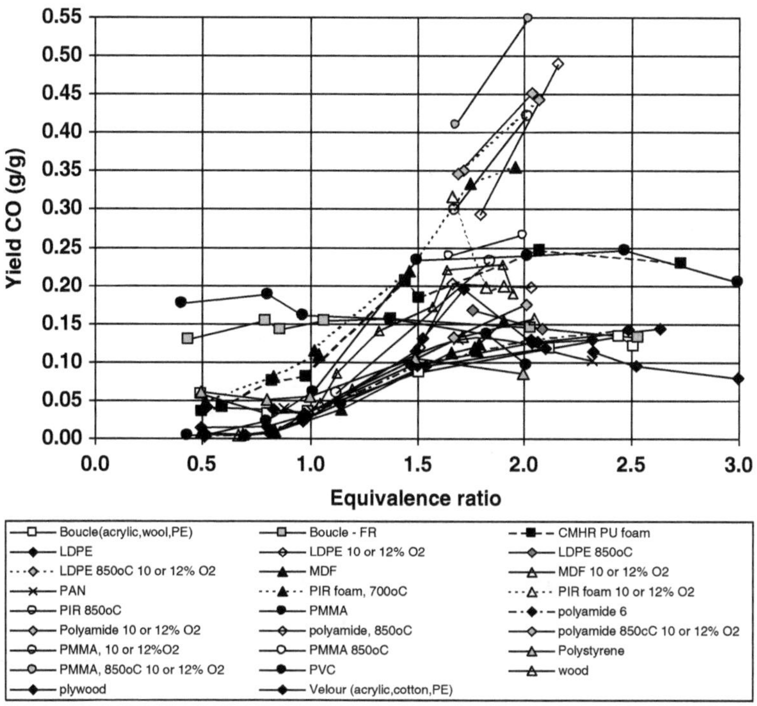

Fig. 2.13 Relationship between equivalence ratio and CO yields measured using the ISO/TS19700 tube furnace [24]

Figure 2.12 shows the relationship between equivalence ratio and effective heats of combustion for a range of common polymers measure using the ISO/TS19700 tube furnace. The effective heats of combustion are maximum under well-ventilated combustion conditions (ϕ 0.5 and 1.0), depending on the polymer composition, but decrease to almost half these levels under fuel-rich condition (ϕ 2.0). For halogenated flame-retarded materials such as polyvinylchloride (PVC) the combustion efficiency is low across the range.

Figures 2.13 and 2.14 show the effects of equivalence ratio on the yields of CO and HCN for the same range of natural and synthetic polymers. For non-flame retarded materials there is a considerable increase in CO yield between well-ventilated and fuel-rich combustion conditions. For example, the CO yield from polymethylmethacrylate (PMMA) increased by a factor of 62 between ϕ 0.5 and 2.0. Under higher temperature (post-flashover, fuel-rich) combustion conditions the CO yields were even higher for some polymers. The presence of halogens as flame-retardants also increased the CO yield under well-ventilated combustion conditions. Figure 2.14 shows similar effects on HCN yields from nitrogen-containing materials, with considerable increases as combustion conditions become fuel-rich. For polyamide-6 the HCN yield increased by factors of between 59 and 101 between ϕ 0.5 and 2.0, depending on temperature and oxygen concentration. It was also found

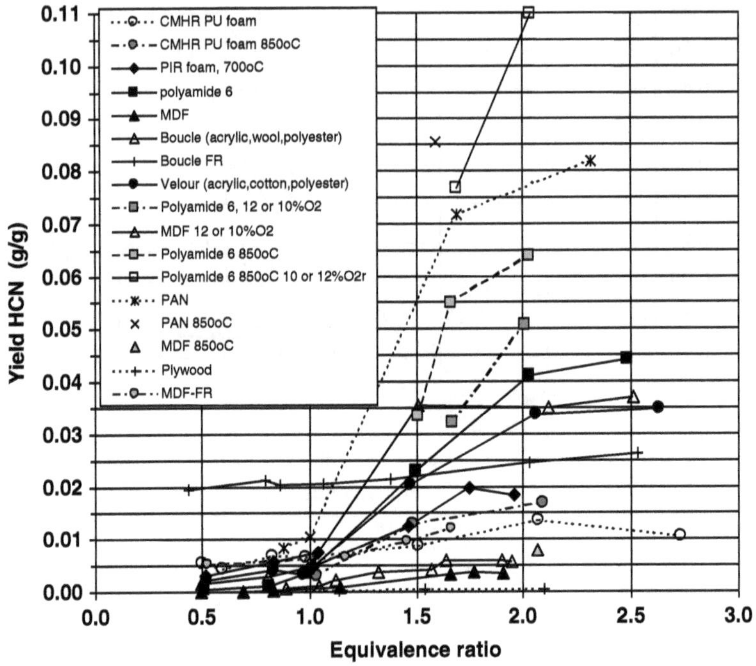

Fig. 2.14 Relationship between equivalence ratio and HCN yields measured using the ISO/TS19700 tube furnace [24, 25]

that there was a close relationship between the efficiency of conversion of fuel carbon to CO and the efficiency of conversion of fuel nitrogen to HCN, which could be useful for modelling applications [25, 26].

The overall findings from this work and complementary studies of full-scale compartment fires and fire investigation statistics is that most flaming fires in buildings are small fires in enclosed dwellings which rapidly become under-ventilated. This results in limited fire damage, usually confined to a limited area of the room of origin, but large volumes of toxic effluents fill open volumes within the structure, resulting in injuries and deaths from toxic smoke exposure. In contrast to this, most large-scale fire tests and fire dynamics models are carried out for well-ventilated combustion conditions involving high heats of combustion and low yields of toxic products. The data obtained using the ISO/TS19700 [16] and the ASTM E2058 flammability apparatus [23], whereby yields are measured as function of equivalence ratio, have been used to derive functions for input to fire dynamics calculations to improve toxic gas concentration calculations for compartment fires, rendering them more realistic and less non-conservative for hazard assessments [1, 24].

2.5 Conclusions

Time-based ASET/RSET analysis is an important and powerful method for evaluating overall fire safety performance.

- For performance-based design, when correctly applied, it enables the engineer to identify and evaluate the effects of each component and their interactions for fire hazard development and fire-safety outcomes.
- It can be applied to scenario analysis with varying levels of sophistication. For example a simple but effective deterministic ASET calculation can be used to determine the tenability time for a single enclosure. A more sophisticated method linked with an evacuation simulation model can be used to calculate the accumulated FED for each individual occupant as they move through building spaces. When such methods are used with variable inputs for different fire and occupant parameters they can be used for Monte Carlo simulation probabilistic risk assessments.
- When the method is applied to the performance-based analysis of prescriptive guidance it can be used to highlight areas requiring modification and provide a means to achieve improvements (pre-empting fire incidents otherwise likely to occur).
- By identifying parameters having a large effect on safety, but for which knowledge is poor, it provides a focus for research needs, the collection of improved data and the development of improved methods.
- It provides a valuable method for incident investigation and forensic fire incident analysis.
- Finally it provides a conceptual framework enabling fire engineers and professionals working on different aspects of fire safety to collaborate to improve overall fire safety knowledge and outcomes.

References

1. Purser, D. A. (2008). Assessment of hazards to occupants from smoke, toxic gases and heat. In P. J. DiNenno (Ed.), *SFPE handbook of fire protection engineering. Chapter 2–6* (4th ed., pp. 96–193). Quincy, MA: National Fire Protection Association.
2. Purser, D.A. (2002). *ASET and RSET: Addressing some issues in relation to occupant behaviour and tenability.* 7th International Symposium on Fire Safety Science (pp. 91–102). Worcester, MA: Worcester Polytechnic Institute. 16–21 June 2002. Fire Safety Science—Proceedings of the Seventh International Symposium. International Association for Fire Safety Science, 2003.
3. British Standards Institution. (2004). *The application of fire safety engineering principles to fire safety design of buildings. Human Factors: Life safety strategies—Occupant evacuation, behaviour and condition* (PD7974-6). London: British Standards Institution.
4. ISO. (2009). *ISO/TR16738 Fire-safety engineering—Technical information on methods for evaluating behaviour and movement of people International Standards Organisation.* Geneva: Author.

5. Reed, R. E. H. (1986). *British fire legislation on means of escape 1774–1974*. BRE Report. Building Research Establishment. Watford, U.K.
6. Joint Committee of the Building Research Board. (1946). *Post-war building studies*. London: The Stationery Office.
7. Communities and Local Government. (2006*). Fire safety: Approved document B* (Vols. 1 and 2). Communities and Local Government. London, The Stationery Office.
8. (2010). *The Building Regulations, England and Wales*. London, The Stationery Office.
9. *Fire statistics in The United Kingdom 2011*. London: Communities and Local Government (published annually). London, The Stationery Office.
10. *Furniture and furnishings (Fire) (Safety) regulations 1988 statutory instrument 1988/No. 1324*. London, The Stationery Office.
11. Babrausakas, V., Fleming, J. M., & Russell, B. D. (2010). RSET/ASET, a flawed concept for fire safety assessment. *Fire and Materials, 34*, 341–355.
12. Bensilum, M, & Purser, D. A. (2002). *GridFlow: An object-oriented building evacuation model combining pre-movement and movement behaviours for performance-based design*. 7th International Symposium on Fire Safety Science (pp. 941–952). Worcester, MA: Worcester Polytechnic Institute. 16–21 June 2002 (Proceedings in press). Fire Safety Science—Proceedings of the Seventh International Symposium. International Association for Fire Safety Science, 2003.
13. Purser, D. A. (2003). Behaviour and travel interactions in emergency situations and data for engineering design. In E. Galea (Ed.)*, Pedestrian and evacuation dynamics 2003: Proceedings of the 2nd International Conference on Pedestrian and Evacuation Dynamics, Greenwich* (pp. 355–370). London: University of Greenwich.
14. Purser, D. A., Fardell, P. J., Rowley, J., Vollam, S., & Bridgeman, B. (1994, January). *An improved tube furnace method for the generation and measurement of toxic combustion products under a wide range of fire conditions*. Proceedings of the 6th International Conference Flame Retardants '94 (pp. 263–274). London: Interscience Communications.
15. ISO/TS 19700. (2007). *Controlled equivalence ratio method for the determination of hazardous components of fire effluents*. Geneva, Switzerland: International Organization for Standardization.
16. Purser, J. A., Purser, D. A., Stec, A. A., Moffatt, C., Hull, T. R., Su, J. Z., et al. (2013). Repeatability and reproducibility of the ISO/TS 19700 steady state tube furnace. *Fire Safety Journal, 55*, 22–34.
17. Purser, D. A. (2011, October 21). *Design behavioural scenarios for escape behaviour modelling in tunnels and underground complexes*. Proceedings of Advanced Research Workshop on Evacuation and Human Behaviour in Emergency Situations (pp. 1–19). Santander, Spain: University of Cantabria.
18. Gwynne, S. M. V., Purser, D. A., Boswelll, D. L., & Sekizawa, A. (2012). Understanding and representing staff pre-warning delay. *Journal of Fire Protection Engineering, 22*, 77–99. doi:10.1177/1042391512436785.
19. Ballard, C. (1996). Düsseldorf Airport tragedy. *Fire International, 152*, 18–21.
20. Rosepark Nursing Home. *Sheriff's inquiry report and appendices*. http://www.scotcourts.gov.uk/opinions/2011FAI18.pdf http://www.scotcourts.gov.uk/opinions/APPENDICES.html
21. Building Research Establishment. (2004, January 31). *Experimental research for Scottish Building Standards Agency following the fire at the Rosepark Care Home,* Glasgow (BRE Project Rep. No. 219132). Retrieved October 2004, fromhttp://www.scotland.gov.uk/Topics/Built-Environment/Building/Building-standards/publications/pubresearch/researchfire/resroprk
22. British Standards Institution. (2003). *PD7974-1 Application of fire safety engineering principles to the design of buildings—Part 1: Initiation and development of fire within the enclosure of origin (Sub-system 1)*. London: British Standards Institution.

23. Tewarson, A. (2008). Generation of heat and gaseous, liquid and solid products in fires. In P. J. DiNenno (Ed.), *SFPE handbook of fire protection engineering* (4th ed., pp. 3-109–3-194). Quincy, MA: National Fire Protection Association.
24. Purser, D.A., & Purser, J. A. (2003, March 28). *The potential for including fire chemistry and toxicity in fire safety engineering* (BRE Rep. No. 202804). Garston: Building Research Establishment.
25. Purser, D. A. & Purser, J. A. (2008). *HCN yields and fate of fuel nitrogen for materials under different combustion conditions in the ISO 19700 tube furnace*. Fire Safety Science—Proceedings of the ninth international symposium (pp. 1117–11128). Karlsruhe, Germany: International Association for Fire Safety Science.
26. Purser, D. A. (2000). Toxic product yield and hazard assessment for fully enclosed design fires involving fire retarded materials. *Polymer International, 47*, 1232–1255.
27. Stec, A., Hull, T.R., Purser, D. & Purser, J. (2014, February 10–14). *Fire toxicity assessment: Comparison of Asphyxiant Yields from laboratory and large scale flaming fires*. 11th International Symposium on Fire Safety Science. Christchurch, New Zealand: University of Canterbury.

Chapter 3
Representation of Human Behaviour in Fire

Steven M.V. Gwynne and Erica D. Kuligowski

3.1 Introduction

For a building to be constructed and occupied, the engineer must first establish that the building affords a sufficient level of safety during a fire incident. Structures are currently designed and constructed in accordance with prescriptive and performance-based (PBD) methodologies to ensure this level of safety.

Prescriptive approaches rely on the application of a predetermined set of rules that, if employed, allow the design to be *deemed* safe [1]. However, the achieved degree of safety is not always apparent. In contrast, performance-based designs rely on a quantitative assessment of the fire and evacuation performance levels achieved. This approach requires the quantification of both ASET (Available Safe Egress Time—the time before conditions become untenable) and RSET (Required Safe Egress Time—the time for the population to get to a place of safety, as represented by the required safe egress time). These are then compared to establish whether there is sufficient time, potentially including a margin of safety, for the population to reach safety before conditions become untenable.

Currently, anywhere from simple engineering equations to complex evacuation modelling techniques are used to estimate RSET for a building or other types of structure.

S.M.V. Gwynne (✉)
National Research Council Canada, Ottawa, Canada
e-mail: smvgwynne@gmail.com

E.D. Kuligowski
Materials and Structural Systems Division, Engineering Laboratory,
National Institute of Standards and Technology, Gaithersburg, MD, USA
e-mail: erica.kuligowski@nist.gov

A. Cuesta et al. (eds.), *Evacuation Modeling Trends*,
DOI 10.1007/978-3-319-20708-7_3

However, these models tend to simplify[1] the behaviour of occupants, if behaviour is simulated at all. Current models that simulate behavioural aspects of evacuee performance place a heavy burden on the practitioner if this is to be done properly. Whereas theories of human movement in evacuation exist [2], and are incorporated into modelling techniques, a comprehensive, generalized theory of human behaviour in fire does not exist.[2]

An understanding of human behaviour in fire when assessing the life safety of a structure is essential. In the past, evacuee decision-making process was assumed to be either panic-based and effectively unpredictable or too complicated to understand and was then excluded from an assessment of performance. These contradictory positions were often employed simultaneously and, although anecdotal examples were identified to support both positions, were largely inaccurate. In the context of this chapter, it is important that we understand evacuee behaviour such that we can model it and use it to quantify performance in the design process. It is certainly impractical (and likely impossible) to predict the actions of any individual with any confidence; however, it is possible to identify the factors that might influence performance and the types of responses associated with these factors. This will then enable us to determine where the decision-making process, as opposed to physical movement, determines the time to reach safety and quantify this process.

The Beverly Hills Supper Club fire [3] provides an example where both of these prior behavioural assumptions are not appropriate and demonstrates the importance of understanding human behaviour in fire. A fire developed in one part of the structure, while several hundred patrons were seated elsewhere, some distance from the seat of the fire. The fire developed and eventually smoke travelled along the adjoining corridor—the corridor used by patrons to enter the structure. The smoke reached the large number of patrons (seated in the Cabaret Room). This was their first notification of the incident and that their familiar means of egress was blocked. There were two emergency exits from the Cabaret Room that were unfamiliar to the patron population and recessed such that they could not easily be seen. A busboy (Walter Bailey) took it upon himself to direct the patrons to both exits. Although over 160 people died in that room, the death toll would have been much higher had the patrons not been made aware of the exits available. This indicates the importance of information to the evacuation process, somewhat contradicting the prevailing wisdom. In addition, Feinberg and Johnson examined the incident to establish the likely occurrence of panic in conditions that would be assumed to promote this behaviour [3]. They postulated that, if panic had occurred, social groups would have broken down leading groups to separate. In contrast, they found that social groups tended to escape or succumb together. This incident provides an example of the importance

[1] This simplification, in particular, the engineering timeline, is discussed in Chap. 1, where the engineering timeline is presented in detail.

[2] Work has been performed by behavioural researchers to develop conceptual models of human behaviour in fire. These models often reflect a specific incident [54] or behavioural from a specific type of building [64]. Work is still required to develop a comprehensive theoretical model for all types of fire scenarios and occupancies/buildings.

of understanding evacuee behaviour (e.g. familiarity, exit awareness, situational awareness, etc.) and the inaccuracy of assuming that evacuee behaviour will be dominated by a panic-based response.

Without a theoretical representation of human behaviour in fire, significant consequences can arise for evacuation model users, evacuation model developers, building owners, public safety officials, codes and standards development organizations, and those who judge evacuation analysis (i.e. the authority having jurisdiction). Not only are model users required to identify/describe the scenario, many current models also require the user to determine the expected behavioural response of the population to some/all of the scenario conditions faced [4]. Evacuation models are also limited in their representation of human behaviour during evacuation; often requiring users to provide a large amount of input data on evacuee behaviours. Input may include what behaviours are performed and potentially the outcome of the performance. In addition, the lack of a generalized theory of human behaviour in fire can result in a greater reliance upon default values/settings provided by the egress model, or an artificial focus on physical responses [5].

What is needed is the development of a comprehensive conceptual model of evacuee decision-making and behaviour during fire events. Such a conceptual model could then be embedded into engineering and computational models to begin to predict human behaviour in fire situations. Insight on human decision-making and behaviour during fires can be gained from disciplines in the social sciences, such as sociology, psychology, and human factors and ergonomics.

To summarize:

- RSET models tend to focus solely on the simulation of human movement, rather than human behaviour, during evacuation.
- Evacuee behaviours can account for a significant portion of overall evacuation time; therefore an accurate representation is essential for many scenarios.
- An understanding of human behaviour is essential when calculating RSET, and the lack of such understanding affects many different stakeholders in the fire protection community.
- A conceptual model of human behaviour in fire is needed.
- Data on human behaviour in fire should be collected to confirm or refine this theory.

3.2 Chapter Purpose

The purpose of this chapter is to identify the ways in which practitioners and model developers can account for human behaviour in fire, both now and in the future, in the use and development of evacuation models. This chapter begins with two sections that discuss our current understanding in human behaviour in fire. Section 3.3 discusses behavioural theory from the natural and technological disasters, not necessarily including building fires, that describes the ways in which occupants make decisions and take protective action during emergencies. Section 3.4 identifies "behavioural statements" or mini-theories from human behaviour in fire studies (or similar). Rather than reflecting a comprehensive theory, these statements provide a piece-meal view of our theoretical understanding of human behaviour during fire.

A discussion of current theory is important because it provides: 1) an indication of our current understanding of evacuee response, which in turn, helps to identify the gaps in our knowledge of the topics, and 2) a benchmark against which current model representation of evacuee behaviour, introduced later in the chapter, can be measured.

The next section, Sect. 3.5, provides an overview of methods that can be used to collect data on human behaviour in fire. This section is important because, without appropriately collected and sufficiently documented data, the development of a comprehensive theory on human behaviour in fire is not possible.

The chapter ends with Sects. 3.6 and 3.7 that focus on representing human behaviour in fire within evacuation models. Section 3.6 discusses the different approaches currently used by evacuation models to represent evacuee behaviour. With the use of these current approaches, it is the hope that practitioners ask questions of their assumptions and of the models deployed in order to produce more credible and robust solutions. Section 3.7 concludes with an example of a conceptual model of human behaviour in fire, intended to get model developers and users thinking of the ways in which to incorporate these types of theories into current evacuation modelling techniques.

3.3 Theory of Human Behaviour in Other Disasters

In a fire emergency, individuals are required to make a concerted effort to create meaning out of new and unfamiliar situations, often under time pressure. From this meaning, a set of actions, different from those that have become routine, must be created. Emergent norm theory (ENT), explains the process of meaning-making in the face of uncertain conditions [6], stating that in situations where an event occurs that creates a normative crisis (i.e. an event where the institutionalized norms may no longer apply), such as a building fire, individuals interact collectively to create an emergent situationally-specific set of norms to guide their future behaviour. In other words, individuals work together to redefine the situation and propose a new set of actions, which is the product of milling and keynoting processes.

Milling is a communication process whereby individuals come together in an attempt to define the situation, propose and adopt new appropriate norms for behaviour, and seek coordinated action to find a solution to the shared problem at hand [7]. The group engages in both physical and verbal communication in order to ask the three following questions: 1) what happened? 2) what should we do? and 3) who should act first? (with the final question referring to leadership selection) [8, 9]. Leaders emerge as keynoters, or those who advance suggested interpretations of the event or suggestions on what do to next [6, 10].

The consequences of the milling process are that individuals become sensitized to one another, that a common mood develops, and that a collective definition of the situation is decided upon that minimizes initial ambiguity [9]. Overall, in the face of new and uncertain situations, milling and the keynoting processes allow the group to define the situation and to propose next steps for alternative schemes of social action [6, 7, 10].

A decision-making model has been developed that extends and applies ENT's explanation of the meaning-making process in crises to disaster situations. The Protective Action Decision Model (PADM), which is based on over 50 years of empirical studies of hazards and disasters [11–15], provides a framework that describes the information flow and decision-making that influences protective actions taken in response to natural and technological disasters [16]. The model posits that cues from the physical environment (e.g. the sight of smoke) or the social environment (i.e. emergency messages or warnings), if perceived as indicating the existence of a threat, can interrupt normal activities of the recipient. Depending upon the perceived characteristics of the threat (e.g. what is going on and how dangerous is it?), individuals will either seek additional information, engage in actions to protect people or property, perform actions to reduce psychological stresses, or resume normal activities (as indicated above) [16]. In addition to perceptions of the threat, responses are also determined by the perceived feasibility of protective actions.

The PADM asserts that the process of decision-making begins when people witness cues from the disaster event. Individuals can encounter only one type of cue (for example, seeing smoke) or may be presented with a variety of different cues, including environmental cues, the behaviour of others, and warning messages. The introduction of these cues initiates a series of pre-decisional processes that must occur in order for the individual to perform protective actions. However, the PADM does not provide detailed information on the *types* of protective actions in which occupants engage and why they engage in these actions during emergencies.

The individual must receive the cue(s). In a fire, occupants are presented with external cues. These cues can be physical or social in nature, meaning that they arise from the physical environment or the social environment; e.g. breaking glass and actions taken by the building population, respectively. These cues can be presented alone or several at a time, depending upon the nature of the event. Physical and social cues produced in a building fire can be received by occupants through hearing (e.g. an alarm or authority warning), smelling (e.g. smoke), seeing (e.g. others running), tasting (e.g. sulphur dioxide or hydrogen chloride), and/or touching (e.g. heat). Given the nature of the situation and individual sensory capabilities, it should not be expected that all people will have access to the same external information or will perceive it in the same way [17]. Just because an individual receives a cue does not necessarily mean that s/he has paid attention to it. Therefore, the next step involves the individual paying attention to the cue(s). This pre-decisional process involves the individual cognitively registering that a cue has been received and beginning to provide the necessary attention, which leads to the last pre-decisional process; i.e. comprehension. Comprehension means understanding the information that is being conveyed. If the message uses a different language or highly technical terms, comprehension will be difficult.

Comprehension also refers to the development of an accurate understanding of environmental cues. For example, will the individual understand that the smoke s/he smells is coming from a building fire rather than from burnt toast in the kitchen?

After the three pre-decisional processes are completed, the core of the decision-making model consists of a series of five questions [16]:

1. Is there a real threat that I need to pay attention to? [If yes, then the individual believes the threat]
2. Do I need to take protective action? [If yes, then the individual decides that s/he needs to take protective action]
3. What can be done to achieve protection? [The individual begins searching for possible protective action strategies]
4. What is the best method of protection? [The individual chooses one of the action strategies developed in the previous stage and develops a protective action strategy or plan]
5. Does protective action need to be taken now? [If yes, the individual follows the plan developed in the previous stage]

Individuals must "answer" each question in order to proceed through the perceptual-behavioural sequence, in which the outcome of the process is the performance of a behavioural action. If, at any time, the individual cannot answer a question, they engage in actions to seek additional information—asking themselves or others: a) what information do I need? b) where and how can I obtain this information?, and c) do I need this information now? A graphic of the process is shown in Fig. 3.1.

After the three pre-decisional processes, the individual first asks "Is there a real threat that I should pay attention to?" In this stage, the individual decides if there is actually something occurring that may require his/her action, sometimes referred to as warning belief [18], but referred to here as *threat belief* to account for people's reactions to all types of environmental cues [16]. If the individual's answer is yes, then s/he is said to believe the threat, and s/he subsequently moves on to consider the next question in the process. Second, a risk assessment is performed. Research has shown that a person's perception of personal risk is highly correlated with disaster response [16]. Here, the individual determines the personal consequences of the threat and asks [19]: "Do I need to take protective action?" At this point, the individual tries to gain insight into the potential outcomes of the disaster and what those potential outcomes mean for his/her safety (i.e. tries to improve their "situation awareness"[20]). The internal dialogue that takes place at this stage can be thought of as mental simulation, in which the individual develops a mental model of what is going on in his/her environment, based on perceived cues, and then expands the mental model to project forward and predict the personal consequences of the event [21]. The more certain, severe, and immediate the risk is perceived to be, the more likely the individual is to perform protective actions [22].

Especially in the initial stages of an event, individuals may have difficulty with the first two questions—identifying and assessing the risk. Even after receiving

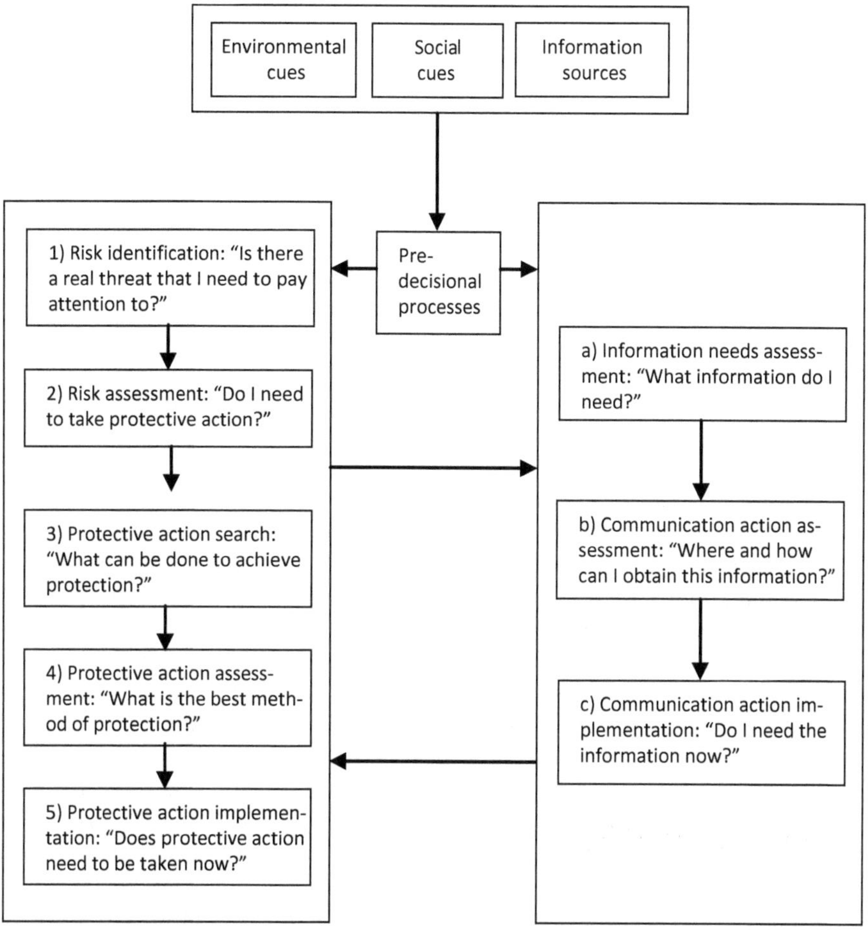

Fig. 3.1 The protective action decision model (Source—[16] redrawn from p. 47)

what many would consider obvious evidence of danger, some people disbelieve or disregard the threat altogether—thinking that nothing unusual is happening that places them at risk, known as normalcy bias [14, 23, 24]. People may also think that even though there may be a threat present, it will not negatively affect them, known as optimistic or optimism bias [25]. Individuals often have trouble estimating the consequences or severity of an incident since they are likely unfamiliar with the potential speed of fire development or lethality of toxic smoke products.

In the third and fourth stages, the individual engages in a decision-making process to identify 1) what can be done to achieve protection; and 2) the best available method of achieving this protection. The outcome of the third stage is a set of possible protective actions from which to choose.

Research literature suggests that occupants develop their options by performing mental simulation [26, 27], similar to the methods of developing interpretations.

Mental simulation [21] allows an occupant to mentally structure scenarios on what s/he would do and how s/he would do it in the current situation—project the current situation into the future and estimate possible outcomes. The search for options becomes the process of mentally developing scenarios of action before actually performing the act.

The search for options of what to do can also occur collectively [6, 9]—either collaboratively or through suggestion by a leader. In addition to interpreting an event, groups work together to plan a coordinated action that will solve the problem presented by the interpretation, if any. Suggestions for actions can come from any member of the group, although leaders are likely to emerge with suggestions of next actions [6, 28]. In the face of uncertainty and time pressure, people are likely to come together, share their interpretations, and define plans for collective action in an event.

Occupants or groups are unlikely to search for a large number of options during the decision-making phase. Research suggests that individuals and groups are likely to develop a narrow range of decision options due to the following conditions: 1) perceived time pressure [29–32]; 2) limited mental resources [33–35]; and/or 3) training and knowledge of procedures [21, 27]. Time pressure, likely in a fire event, causes occupants to perceive a fewer number of cues (termed perceptual narrowing), process the information less thoroughly and in turn, to consider a narrow set of options [29]. Also, people do not expend large amounts of intellectual resources envisioning the broad range of scenarios, but rather are likely to envision only the scenarios that they believe are necessary to reach a goal [34]. Finally, research suggests that occupants who are highly trained and/or know of specific procedures will be guided by training and will likely not develop more than one option at a time [21].

The fourth stage involves protective action assessment. This stage involves assessment of the potential option(s), evaluating the option(s) in comparison with taking no action and continuing with normal activities, and then selecting the best method of protective action.

Rationality-based research claims that occupants will attempt to optimize their decision-making by considering all options developed and choosing the best one—known as rational choice strategy [36, 37]. In a fire situation, weighing of multiple options is unlikely to occur. Research on decision-making under uncertainty indicates that occupants use a variety of heuristics to make this choice [21, 38]. Heuristics are simple rules to explain how individuals make decisions. Whereas some research might view the use of heuristics as a source of bias in decision-making [39], other researchers see heuristics as strengths based on the use of expertise and experience [40]. Examples of heuristics that occupants employ in choosing options include anchoring or focusing on the first option developed [38], choosing the most available option (the easiest to develop or recall) [38], comparing all options with each other and choosing one based on the evaluation criteria [41–43], and satisficing [33].

Satisficing [34, 44] is a method in which an individual chooses the first option that seems to work, though not necessarily the *best* option overall [21]—an option which produces results that are good enough rather than optimal. The satisficing heuristic actually combines the processes of option development and option choice together in one step. As the decision-maker develops options, s/he evaluates each one as it is developed and stops developing options when one is deemed to satisfy the search criteria.

Whereas the rational choice strategy is more likely to be used when people attempt to optimize a decision [21], satisficing is more likely to be used in situations with a greater time pressure, dynamic conditions, and ill-defined goals [21].

In emergencies, individuals at risk have two general options: taking protective action or continuing previous activities. Once an action is chosen, the end result of stage 4 is an adaptive plan, which can vary in its specificity. For example, for households under disaster threat conditions, an adaptive plan can include a specific evacuation plan that outlines the travel destination (e.g. a relative's house), a route of travel to that destination, and a means of getting there (i.e. some form of transportation).

After a protective action is chosen and the adaptive plan is developed, occupants *may* enter stage 5 and perform the action that they decided upon in the decision-making phase. If new information is presented before an action is performed, the occupant will reconsider and discard the current action and begin the behavioural process again. The action involves performing some type of physical act although the act could be waiting, or even inaction, that takes some amount of time to complete (or is conducted for a period of time). Both summary research (e.g. [45–47]) and research on specific incidents (e.g. [48–51]) highlight certain actions in which occupants are likely to engage [52].

These protective actions, depending upon the situation, can include waiting, alerting others, preparing for evacuation, assisting others, fighting the fire, and searching for and rescuing others. However, if information received is incomplete, ambiguous, contradictory, or causes uncertainty in understanding cues and which actions to take, then, individuals will likely engage in additional information-seeking actions (shown by the right-hand column in Fig. 3.1). These can include milling, physically seeking information, and/or asking others for information. The greater the ambiguity perceived, the more likely that individuals will search for additional information that can guide their actions [6, 53, 54]. Any information gained will then act as social or physical cues to begin the decision-making process over again.

Note that individuals do not have to go through each stage or question in the decision flow chart shown in Fig. 3.1. For example, if an individual is presented with information about the event from a credible source or if s/he is ordered to evacuate, s/he may move on to later stages in the decision process rather than going through each one in succession. Finally, individuals who decide that they are not at risk, may neglect to take protective action at all and in turn, terminate the emergency decision-making process.

3.4 Behavioural Statements of Human Behaviour in Fire

Section 3.3 described the process through which individuals make protective action decisions in response to fires and other types of disasters based on the PADM [16]. However, the PADM, and other supporting models, do not provide sufficient information on the specifics or the types of protective actions in which occupants engage and why they engage in these actions during emergencies, which is necessary for the development of a comprehensive theory of human behaviour in fire.

Given the absence of a comprehensive human behaviour in fire theory, the model users and developers are faced with the task of assembling disparate pieces of knowledge from available sources in the field or beyond. Data have been collected on human behaviour in fire; however, it is typically distributed across different resources, presented in different formats and placed in different contexts—producing an incomplete, disorganized, and disparate understanding of the subject matter. In this section, key human behavioural statements currently available for fires and other emergencies are presented and discussed. This discussion is intended to form a basis from which a practitioner may inform their work.

24 behavioural statements are presented here, derived from a variety of sources, to describe human behaviour in fire [45]. Even though the statements presented do not represent a complete picture of human behaviour during fire, they do provide important information about the following:

- behaviours that may occur during a building evacuation,
- factors that influence these behaviours,
- possible outcome of these behaviours (i.e. significant delay times).

The statements presented below are typically derived from incidents, (repeated) observations or aspects of existing theories in adjacent fields that have been co-opted into evacuation analysis. This list is by no means exhaustive, but represents the key behavioural conventions that are identified, understood, and employed within models and engineering practice.

In this section, the behavioural statements are organized to aid in both the understanding and location of the statements within the decision-making process.

3.4.1 [Phase 1] Perceiving or Receiving Cues and Information

1) Content of the cue matters. The precision, credibility, clarity, comprehensiveness, intensity and specificity of the external cues will affect the assessment of the information in the individual's decision-making process [19].
2) Authority of the information source affects the perceived credibility of the information [55].
3) The actions of the surrounding population can influence the internal processes and the actions of the individual; e.g. the use of routes/space by others increases their attractiveness [56].
4) Some individuals exhibit hypervigilance that makes them particularly sensitive to certain cues [52].
5) Previous experience of false alarms or frequent drills can reduce sensitivity to alarm signal [57].
6) Habituation (where a process has become routine in nature), focus and stress can narrow the perceptual field, and thus, not all available cues will be internalized [58].
7) Sensory impairments can inhibit the perception of cues [59].

3.4.2 [Phase 2] Assessing the Situation and Perceiving Some Level of Risk

8) Normalcy bias and optimism bias are commonplace. In other words, people often think that nothing serious is taking place, and that nothing bad will happen to them, respectively [24, 60].
9) Training may allow the incident to be defined more quickly by the evacuee and provide pre-determined viable responses [21].

3.4.3 [Phase 3] Selecting a Response or Action

10) People tend to satisfice rather than optimize. In other words, they are more likely to choose an option that is perceived as "good enough" rather than the best option [33].
11) Presence of smoke does not always preclude the use of a route [45].
12) Training and experience may increase an individual's familiarity with the use of components/devices and subsequently improve their use [46].
13) Pre-event commitment to a particular activity may cause individuals to decide against taking protective action [61].

3.4.4 [Phase 4] Influencing Action Selection

14) People have different abilities that influence action selection [59].
15) People seek information in situations where information is lacking or incomplete [45, 46, 52].
16) People engage in protective actions, including preparing to move to safety or helping to protect others from harm, before they initiate a movement towards safety [45, 47, 62, 63].
17) People move towards the familiar, such as other people, places and things [61].
18) People may re-enter a structure, especially if there is an emotional attachment to the structure, the contents and/or the inhabitants [45].

3.4.5 [Phase 5] Influencing the Overall Decision-Making Process

19) People will behave in a rational AND altruistic manner; panic is rare [45, 64, 65].
20) Uncertainty, time pressure and volume of information can increase stress levels [30, 34].

21) Pre-incident experience influences how cues are processed, how the situation is defined and how protective actions are selected [52].
22) Evacuation is a social process, in that groups are likely to form during an evacuation [66].
23) Social rules and roles[3] in place prior to a fire event form the basis of those employed during the event. In other words, people's role before the incident will influence their performance during the event [8].
24) New norms may emerge where existing normative structure is incapable of addressing the new fire situation [6, 7].

The 24 behavioural statements just discussed represent qualitative mini-theories from over 50 years of research of human behaviour in fire and other emergencies. These statements provide practitioners with helpful guidance as to the elements that will need to be represented within an evacuation scenario when calculating RSET for a building or other type of structure. However, at present, these statements are disconnected, making it difficult for practitioners to account for human behaviour in fire when performing egress calculations and even more vital to develop a functioning, comprehensive conceptual model of human behaviour in fire, that will improve the techniques currently used to evaluate evacuation timing. Once a comprehensive conceptual model is developed, these statements will likely form the basis of a validation case against which a new conceptual model might be compared.

As mentioned previously, it is unlikely that we will be able to produce a conceptual model that can identify any specific individual's response with a high degree of confidence. However, we do need a model that can identify the process that an individual goes through, connecting the factors influencing performance to the actions forming evacuee response. Undoubtedly, some of the elements of the process will be better understood than others. Also, the eventual response selection may hinge on arbitrary/undocumented/coincidental factors that are unknowable and not represented in the model; i.e. the same individual may respond differently in similar situations. Given this, a stochastic element will likely be introduced to compensate for the lack of refinement of the model and the many known (and unknown) factors that might influence the outcome—both as a recognition of the limitations of any conceptual model and that scenarios will include factors that are novel and that fall outside of the scope of the model.

3.5 Data Collection

The first step in the development of a fire-related behavioural conceptual model is data collection of human behaviour in fire data, which has been limited to date. Human behaviour in fire is a relatively new field that has emerged within the last few decades. Prior to this time, human response was assumed to be dominated by

[3] Influenced by the peer group, the organization, and/or broader culture/social entities.

physical factors, to be panic-based, considered intractable, and/or was excluded from engineering practice entirely. A considerable effort has been required to convince practitioners of the need to understand and address human behaviour in fire within engineering analysis—it must be included in design solutions. As highlighted in the introductory section, recording the needs and responses of the evacuee can influence the time it takes for them to initiate their response, the routes adopted and then the time for them to reach a safe place (e.g. Beverly Hills Supper Club [51]).

Data-sets are the foundation of any comprehensive conceptual model or theory of human behaviour in fire, which can then support the tools employed in the field and inform practice [67, 68]. Along with theory, data also plays an important role in validation exercises, to ensure that the conceptual model developed represents an appropriate picture of real-world settings.

Behavioural data-sets are difficult and expensive to gather [70]. This is because typical laboratory experiments (which are difficult enough to conduct) provide an artificial environment that will influence the relevance of the results produced by a test—therefore, real-world situations are required that are more costly, less controlled, and more ethically challenging. Therefore, it is important to optimize the collection and application processes. In other words, data collectors should be working to obtain data that are relevant and necessary for the development of the conceptual model, as well as provide the context under which the data were collected to ensure its appropriate use.

Behavioural data can be derived from an array of different fire-related scenarios, shown in Table 3.1. An understanding of these scenarios provides the context under which these data were collected. It is essential for the researcher to report the source of the data, as well as the data collection and analysis methods, to allow a third-party user to determine appropriateness to the intended applications. Table 3.1 describes the scenarios from which data can be collected, including routine conditions, exercises or drills, experiments, actual incidents, and simulation studies.

In addition to understanding the scenarios under which data can be collected, it is also important to understand the methods that can be employed to collect the data. There are a number of methods available to collect data, a selection of which is described in Table 3.2. In Table 3.2, some examples of the types of data that can be collected using each method are provided along with whether the method can be applied during or after the incident (*Timing*). Each of the methods has their strengths and weaknesses, depending on the resources, access, and prior warning available to a researcher for a particular scenario.

Once collected, in most instances, only a sub-set of the data collected is shared. Often, data are shared in a reduced or distilled format, rather than in a complete format. Potentially more importantly, in the vast majority of cases, only a limited amount of information is provided on the background conditions evident during the original event; i.e. the event described by the data. The reduced dataset and limited context requires a greater degree of interpretation by the third party. This increases the potential for the underlying causal factors being misunderstood, the results being misinterpreted, and the dataset being inappropriately applied.

Table 3.1 Scenarios from which data can be collected

Scenario	Description
Routine conditions	Non-emergency situations, such as general circulation or routine building services/operations
Exercises/drills	Non-emergency evacuation exercises from a structure, typically as part of the safety training/testing schedule. Occupants are evacuated in accordance with the building's emergency procedure as if an incident has occurred. Prior knowledge/warning is a key consideration, since the building management, the evacuating population and/or fire marshals may have been forewarned
Experiments	An attempt to control for specific conditions of interest. The scenario would need to be managed by the researcher to ensure that the actions investigated were performed under representative conditions. This might involve the use of set-piece laboratory trials or the use of virtual reality or gaming systems to collect information on what individuals might do in a particular (virtual) situation
Actual incidents	An actual evacuation from a real fire event. In this scenario, the researcher has no control over the scenario or the conditions present during the evacuation
Simulation	The generation of evacuee responses given input scenario conditions by an evacuation simulation model. Simulation enables the researcher to control the scenario (s) modelled, according to the functionality of the model[a]

[a]In some instances, a simulation is considered a controlled numerical experiment—where specific conditions are tested according to an experimental model, with the limitations of the model outlined, allowing the credibility of the results to be reasonably assessed. This approach may allow for conditions to be examined that might otherwise be beyond reach and also attributes to be controlled, allowing greater confidence in the scenario examined. The major question here, as elsewhere, would be the validity of the experimental model

Therefore, the authors of this chapter suggest the standardization of data collection, analysis, and reporting of results in the field of human behaviour in fire. A need for such a standard tool—at least in the arena of measuring risk perception in fire and disaster events—has already been established in other publications [69, 70]. Similar efforts have already been developed for the collection and reporting of data collected for evacuation movement [67], and such efforts should be developed and followed in the area of human behaviour in fire. This will not only aid in the development of the conceptual model, but also the larger-scale use of these data for various purposes within the fire safety discipline.

3.6 Model Representation

Egress models are an important tool—they allow evacuee performance to be quantified and simulation experiments to be performed to gain insight into the impact of scenario conditions upon the outcome of an evacuation. Egress computer simulation

Table 3.2 Data collection methods

Data collection method	Examples of data	Timing
Video camera: A stream of information is collected allowing action/event processes to be observed and recorded	Velocity, flow, chains of events, actions, etc.	During event
Still camera: A snapshot of an event is recorded allowing specific conditions at a particular time to be recorded, rather than the process by which they developed	Conditions at a point in time; e.g. spacing between people in a crowd, location and size of a queue at a point in time, etc	During Event
Manual observations: Observer records events, conditions or processes to a relatively crude degree of accuracy	Existence of certain conditions, manual timing of arrivals	Dynamic or static/ during event
Survey/Interview: Individual experiences established (either directly or from reviewing reported events/data-sets [51] via secondary material) to build an understanding of the incident	The "hows" and "whys", e.g. decisions to initiate evacuation, specific conditions experienced, actions taken, previous experiences/history, etc	After event
Sensor: Performs pre-defined, automated observations to record specific events over a period of time	Arrival of evacuees at a pre-determined location (e.g. timing), biometrical measurements, physiological measurements, etc.	Static
Tagging: Performs pre-defined, automated observations to record movement of specific individual(s) over a period of time	Arrival of evacuees each equipped with a sensor to denote their movement, etc	During Event

models adopt an array of different approaches to represent evacuee behaviour (i.e. behavioural data or statements). Of most relevance here, in this chapter, is how the models represent individual evacuee response. The current approaches of egress models in representing human behaviour in fire are discussed along with their implications for practitioners.

The authors have reviewed many of the evacuation models currently available, including components that contribute to the overall effectiveness of the model [71, 72]. In this section, the components that relate specifically to simulated behaviour are discussed.

As mentioned in Sect. 3.1 of this chapter, current evacuation simulation models are limited in their capabilities to represent human behaviour in fire. This limitation manifests in two ways: (1) some behaviours are represented, but in a partial manner; (2) other behaviours are excluded entirely.

The behaviour of the population is currently represented by evacuation models in a number of different ways, but can be grouped into five categories:

In a minority of models, especially the older models, there is *no representation of behaviour*. Only the movement of the evacuating population is simulated with no local decision-making processes represented. These models are often composed of a series of engineering calculations, where an evacuating population is treated like a laminar flow, unless the engineer manually represents evacuee behaviour within the scenario [71, 72]. For example, the user may remove an exit from the building configuration (i.e. an unfamiliar exit), and then calculate the time that evacuees might take if they used other available exits (i.e. the main or more familiar exit).

Most evacuation models represent behaviour *implicitly* by assigning certain response delays or occupant characteristics that affect movement throughout the evacuation and are deemed to take into account the delays that expected behaviours might have incurred. Should the explicit representation of behaviours be required in implicit models, it would be based on user input. The user input specifically represents behavioural performance (e.g. travel speed, route use, etc.) to be assessed by the model, rather than allowing the model to predict whether the action will occur at all. Therefore, behavioural actions are an input rather than an output. The process requires the practitioner to develop representative scenarios *and* impose credible evacuee responses. Both of these activities require informed engineering judgment.

Other (agent-based) models represent behaviour in a *conditional* manner, where the evacuee responses to certain conditions are pre-defined, such that individual responses are based entirely on local conditions. For example, structural or environmental conditions of the evacuation would influence evacuee response, possibly by influencing his/her route choice. Models tend to account for conditional behaviours using "if, then" statements or rules. Depending on the sophistication of the approach, this may be entirely deterministic or probabilistic. Additionally, behavioural statements or rules can be assigned based purely on random variables, defined here as *stochastic* behavioural modelling.

Conditional and stochastic approaches require the model to *generate* the evacuee response directly given the conditions faced. Current models often apply approaches derived from other physical sciences (representation through analogy) to describe the movement and interaction of the evacuating population, including approaches representing particle flow, granular flow, potential/magnetic fields [72]. Behaviours are represented on a stimulus-response basis (e.g. cue leads directly to action), rather than representing the complex cognitive, social and adaptive decision-making processes involved in predicting response (i.e. the PADM).

Finally, a few models represent behaviour in an *adaptive* manner, where behaviour is determined according to the experiences and knowledge of the simulated evacuee and the local environmental conditions. The adaptive method reflects an attempt to represent the adaptive capacity of the decision-making process. The user configures the implemented model to represent the scenario conditions. The simulated evacuee is then sensitive to conditions that influence the decision-making process (potentially in conjunction with existing internal information) and, in turn, an action is selected. This approach is:

- Sensitive to the impact of external and internal conditions,
- Establishes a response as the result of a process
- Produces conditions that emerge as a result of interactions between evacuees and their environment.

Currently, no model is entirely adaptive. Instead, some aspects of evacuee behaviour might be represented in an adaptive manner. For instance, exits may be selected by agents' assessment of the impact of the congestion at each exit or the presence of smoke along the route upon their evacuation.

In reality, current models adopt a range of these approaches for different aspects of the evacuation process—with an overall composite approach adopted. For example, an agent may start to evacuate after sensing smoke (*conditional*), move towards an exit according to a speed-density relationship (*implicit*) and then select an exit according to the conditions faced and the assessed impact on his/her performance (*adaptive*). This composite approach makes it difficult for the user both to interpret the validity of model and the results produced.

To improve the current state of evacuation modelling there is a need to implement a comprehensive conceptual model that more accurately represents the decision-making process. This improvement would allow models to represent agent response, which would, ideally, address a broader array of initial conditions and allow for a more comprehensive set of responses. This development would need to be accompanied by a similarly comprehensive validation effort. A validated partial implementation ensures confidence in an incomplete representation of evacuee performance; an unvalidated comprehensive implementation provides a model with sufficient scope, but whose results may not be credible for the scenario of interest.

3.7 Conceptual Model Development

There is a need for the development of a conceptual model of evacuee behaviour. This will rely on data and theory to be embedded in an evacuation model to more credibly predict evacuee actions [52, 68–76]. There has been work done in this area, upon which the field can build; however, none of this work has been fully validated or covers sufficient subject areas to enable a comprehensive conceptual model to be fully developed and implemented.

It should be noted that much of this work pre-dated the use of evacuation simulation tools as a means to quantify performance. Therefore, the insights provided by these research efforts, while important, are not always easily translated into a format that can then be implemented within a computational tool; at least not without significant effort.

An early set of conceptual models in the fire field was developed by Canter, Breaux, and Sime [73]. They developed decomposition diagrams for various types

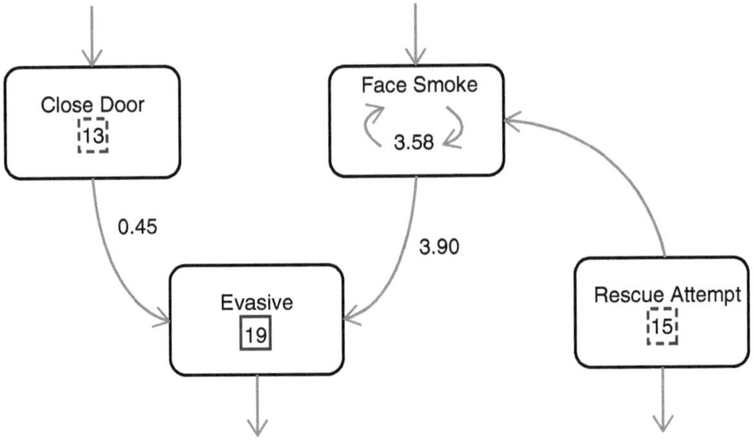

Fig. 3.2 Section of decomposition diagram from Breaux et al. [73]. *Arrows* indicate association; linked probabilities indicated strength of associated. Reworked from [73]

of fire events that identify the sequence of actions performed (see Fig. 3.2). Critically, these diagrams charted the actions performed and the relationship between them; i.e. the probability of a particular action preceding or succeeding another action. Their UK study examined 14 domestic fires, eight multiple-occupancy fires, and six hospital fires, examining the actions of 198 evacuees in total. Interviewees were first asked to give a detailed account of their experience, after which interviewers focused in on certain issues, including recognition of the fire event, location of the occupant, ongoing behaviour, sequence of actions, perception of the situation, past experiences, and background information.

Canter, Breaux and Sime's decomposition diagrams begin to tie various sub-theories together, but focus primarily on the linking of evacuation actions together. Taking this further, Kuligowski developed a model that identifies the interpretations and levels of risk perception that are influential to occupant's actions—specifically for the 2001 World Trade Center (WTC) Disaster [52]. Kuligowski's model is a qualitative model that predicts the pre-evacuation actions of survivors from the WTC incident. The research involved analysing transcripts (derived from the HEED database [77]) from 245 face-to-face interviews with survivors from both WTC towers. The goal of this research was to describe evacuation decision processes in greater detail than either research on building fires or studies on community-wide evacuation, focusing on how people perceive and interpret environmental cues and warnings, how they seek confirmation during sense-making and milling processes, and what they do before moving to safety.

Kuligowski's research produced five main findings:

- The WTC pre-evacuation period was divided into two main phases: the milling/sense making phase (where occupants either continued to work or sought information) and the protective action phase (where occupants engaged in actions that were focused specifically on protecting themselves or others). Both phases took place before the evacuees moved to the stairs or elevators.

- Risk perception, or the feeling of personal danger, was the main predictor of when individuals decided to evacuate. Both individual and environmental factors were identified as influential of risk perception development.
- Some individuals decided to evacuate relatively early. These were primarily higher-level managers, fire wardens, military personnel, or individuals who had experiences with or occupations in emergency situations. These individuals still required information that increased their level of perceived risk. However, they were more inclined to act before others because they felt responsibility for others and/or had previously experienced/witnessed negative consequences associated with fire or building evacuations.
- Certain factors, such as personal responsibility, social connections, and the actions of others, influenced which protective actions people engaged.

Kuligowski's model is not without limitations: (1) it focuses specifically on the pre-evacuation period of one building event; (2) the model does not incorporate any decisions or actions of the decedents; and (3) the factors that influenced each action performed were specific to an office building fire and subsequent evacuation, thus making it difficult to generalize the findings. It is then recognized that this work is a first start to developing a model to predict actions taken during building fires. This research should be expanded upon to include findings from analysis of other building fires, including fires in different types of structures and with different populations, as well as from analysis of other types of disasters, not limited to building fires.

Kuligowski's research was conducted during the current era when simulation tools are employed. Given this, Gwynne examined her research (and the underlying research on which it is based) to identify the types of structures that would need to be included within a model to reflect the elements of Kuligowski's conceptual model [68]. A simple schematic of this translation is shown in Fig. 3.3. This is a modest development in the creation of a conceptual model that is implementation ready; however, it may help facilitate a discussion between researchers developing conceptual models and model developers who might implement them.

After the model specification, a key step in the transition from a comprehensive conceptual model to the implementation of such a model is to predict the timing associated with the performance of certain actions. The extent to which the simulated environment will reflect the consequences of evacuee actions is beyond this discussion; however, the time taken to select and perform actions is critical in assessing the overall time to reach safety. The quantification process will allow the impact of specific actions on overall performance to be determined.

There have been several studies that attempted to predict how long people delay before evacuating [45, 46, 53], the constituent parts of this delay (e.g. [74]) as well as the time it takes individuals to evacuate via stairs [75], etc. This type of research would need to be compiled to support and quantify the decisions and actions suggested by the conceptual model. These decisions/actions (and resultant time delays) would be recombined depending on the decisions made by the individual in question

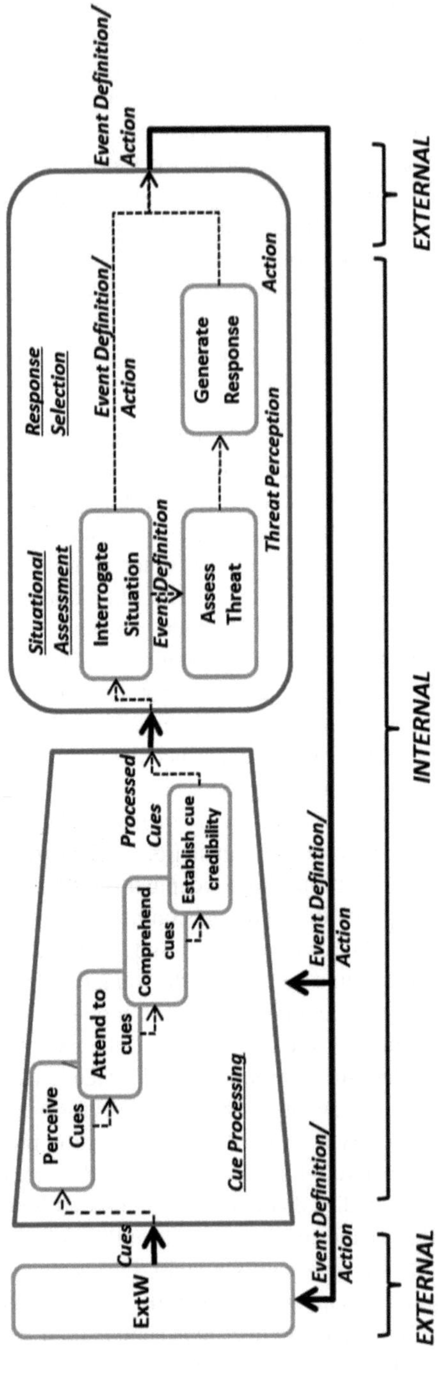

Fig. 3.3 Outline for a conceptual model [68]

to quantify their performance. Efforts have also been made at compiling empirical models based on existing data to reflect elements of the evacuation process (e.g. [76]). With due care, these types of models might be embedded within a comprehensive conceptual model to account for the delays associated with specific tasks, as opposed to determining the tasks actually conducted.

While these models do provide a path forward to identifying the methods that could be used in its eventual development, there are still significant conceptual and empirical gaps. There is much work still to be done to improve our understanding of human behaviour in fire; without this understanding, a comprehensive model is not possible. Listed here are just a few examples of areas in the field that require further study to enable us to better model evacuation from building fires:

- The influence of fire's toxic products and heat on decision-making and behaviour (before incapacitation or death occurs).
- An identification of all of the factors that influence risk perception and how they interact to increase or decrease risk perception levels.
- The types of protective actions that are performed.
- The factors that influence the various types of protective actions performed.
- The factors that influence the receipt of cues, the ways in which people pay attention to cues, and the comprehension of cues.
- The ways in which individual factors, such as gender, disability, age, body size, culture, marital status, past experiences, training and social role, influence decision-making.
- The timing associated with the performance of behaviour and the factors that influence this timing.
- The influence of urgency or other types of dissemination techniques on the response of building occupants.
- The influence of group dynamics on individual decision-making and group decision-making.
- The role of place (including building type or building characteristics) on decision-making.
- The role of psychological states, including stress or anxiety, on decision-making.

For the field to reach its goal and develop a larger understanding of human behaviour in fire, accurate, rigorous, and comprehensive research must continue. Thankfully, there is an array of researchers who are engaged in addressing many of the questions listed and subsequently the furtherance of our understanding of evacuee decision-making and performance.

3.8 Concluding Remarks

The purpose of this chapter is to identify the ways in which practitioners and model developers can account for human behaviour in fire in the use and development and application of evacuation models, both now and in the future. Human behaviour in fire is an important part of the RSET calculation; however, not much in the way of guidance is provided to the practitioner, given the limited and disparate data available and the lack of a comprehensive conceptual model describing evacuee response. Given this, the inclusion of it within the modelling process is often reliant upon the capacity of the user to manually represent evacuee behaviour. Hopefully, this chapter has increased this capacity by encouraging the practitioner to scrutinize the theory, data, models and engineering practice employed in order to ensure the reliability and credibility of any assumptions made and any tools applied. Practitioners are often faced with the difficult task of identifying scenarios of interest, representing them within models and then assessing the results. Similarly model developers are faced with the task of assessing disparate data-sets and theories, and then combining them into a useful tool. Both of these tasks are extremely challenging and the efforts of these groups should not be underestimated. It is hoped that by identifying the key challenges faced that this chapter helps further the conversation between researchers, practitioners and developers to enhance current understanding and ensure that it is shared and employed appropriately.

Looking ahead, to enhance the representation of the simulated evacuee, more representative data-sets should be collected, described in a more consistent and comprehensive manner (ideally in relation to a comprehensive theory of human behaviour in fire) and made available to model developers and practitioners. None of these developments are trivial. However, there is a growing consensus that simulation models are and will remain a key tool in quantifying safety and a tool to help impact life safety when in the hands of informed and expert practitioners. Suitably enhanced tools would then better equip the engineering community to forecast and assess evacuee performance.

References

1. Communities and Local Government. (2006). *Approved document B, Vol. 1.: DwellingHouses*, Building Regulations 2000. London: NBS.
2. Gwynne, S. M. V., & Rosenbaum, E. R. (2009). Employing the hydraulic model in assessing emergency movement. In P. DiNenno et al. (Eds.), *The SFPE handbook of fire protection engineering* (4th ed., pp. 3-373–3-396). Quincy, MA: National Fire Protection Association.
3. Feinberg, W. E., & Johnson, N. R. (2001). The ties that bind: A macro-level approach to panic. *International Journal of Mass Emergencies and Disasters, 19*(3), 269–295.
4. Society of Fire Protection Engineers. (2001). *Engineering guide to human behaviour in fire: Review draft*. Society of Fire Protection Engineers, MD, USA.
5. Gwynne, S., Kuligowski, E. D., & Spearpoint, M. (2012, September 19–21). More faults with defaults. *Proceedings of the 5th International Symposium, Human Behaviour in Fire 2012*

(pp. 263–274). Cambridge, England: Interscience Communications Ltd. ISBN 978-0-9556548-8-6.
6. Turner, R. H., & Killian, L. M. (1987). *Collective behaviour*. Englewood Cliffs, NJ: Prentice Hall.
7. Aguirre, B. E., Wenger, D., & Vigo, G. (1998). A test of the emergent norm theory of collective behaviour. *Sociological Forum, 13*(2), 301–320.
8. McPhail, C. (1991). *The myth of the madding crowd.* New York: Walter de Gruyter.
9. Turner, R. H., & Killian, L. M. (1972). *Collective behaviour*. Englewood Cliffs, NJ: Prentice Hall.
10. Connell, R. (2001). *Collective behaviour in the September 11, 2001 evacuation of the World Trade Center* (Preliminary paper #313). Newark, DE: University of Delaware Disaster Research Center.
11. Sorensen, J. H., & Vogt-Sorenson, B. (2006). Community processes: Warning and evacuation. In H. Rodriguez, E. L. Quarantelli, & R. R. Dynespp (Eds.), *Handbook of disaster research* (pp. 183–199). New York: Springer.
12. Mileti, D. S., & Peek, L. (2001). Hazards and sustainable development in the United States. *International Journal of Risk Management, 3*(1), 61–70.
13. Tierney, K. J., Lindell, M. K., & Perry, R. W. (2001). *Facing the unexpected: Disaster preparedness and response in the United States*. Washington, DC: Joseph Henry Press.
14. Drabek, T. E. (1986). *Human system responses to disaster: An inventory of sociological findings*. New York: Springer.
15. Mileti, D. S., Drabek, T. E., & Haas, J. E. (1975). *Human systems in extreme environments: A sociological perspective*. Boulder, CO: Institute of Behavioural Science, University of Colorado.
16. Lindell, M. K., & Perry, R. W. (2004). *Communicating environmental risk in multiethnic communities*. Thousand Oaks, CA: Sage Publications.
17. Gwynne, S. M. V. (2007). *Optimizing fire alarm notification for high risk groups: Notification effectiveness for large groups*. Summary Report, prepared for The Fire Protection Research Foundation, NFPA, Quincy, MA.
18. Mileti, D. S. (1974). *A normative causal model analysis of disaster warning response*. Boulder, CO: University of Colorado Department of Sociology.
19. Mileti, D. S., & Sorensen, J. H. (1990). *Communication of emergency public warnings: A social science perspective and state-of-the-art assessment.* Oak Ridge TN: Oak Ridge National Laboratory, U.S. Department of Energy.
20. Groner, N. (2009). *A situation awareness analysis for the use of elevators during fire emergencies*. In 4th International Symposium on Human Behaviour in Fire: Conference Proceedings (pp. 61–72). London, England: Interscience Communications.
21. Klein, G. (1999). *Sources of power: How people make decisions*. Cambridge, MA: The MIT Press.
22. Perry, R. W., Lindell, M. K., & Greene, M. R. (1981). *Evacuation planning in emergency management.* Lexington, KY: Lexington Books.
23. Sekizawa, A. (2004). Human factors in fire safety measures. *Fire Science and Technology, 23*(4), 296–302.
24. Okabe, K., & Mikami, S. (1982). *A study on the socio-psychological effect of a false warning of the Tokai earthquake in Japan.* Paper presented at the Tenth World Congress of Sociology, Mexico City, Mexico.
25. Kunreuther, H. (1991). *A conceptual framework for managing low probability events.* Philadelphia: Center for Risk and Decision Processes, University of Pennsylvania.
26. Gwynne, S., Galea, E. R., Lawrence, P. J., Owen, M., & Filippidis, L. (1999). Adaptive decision-making in response to crowd formation in building EXODUS. *Journal of Applied Fire Science, 8*, 265–289.
27. Thompson, C. M., Forester, J. A., Cooper, S. E., Bley, D. C., & Wreathall, J. (1997). *The application of ATHEANA: A technique for human error analysis*. Proceedings of the 1997 IEEE

Sixth Annual Conference on Human Factors and Power Plants (pp. 9-13–9-17). Orlando, FL: IEEE.
28. Connell, R. (2001). *Collective behavior in the September 11, 2001 evacuation of the World Trade Center* (Preliminary Paper #313). Newark, DE: University of Delaware Disaster Research Center.
29. Karau, S. J., & Kelly, J. R. (1992). The effects of time scarcity and time abundance on group performance quality and interaction process. *Journal of Experimental Social Psychology, 28*(6), 542–571.
30. Zakay, D. (1993). The impact of time perception processes on decision making under time stress. In O. Svenson & A. John Maule (Eds.), *Time pressure and stress in human judgment and decision making* (pp. 59–72). New York: Plenum Press.
31. Janis, I. L. (1982). Decision making under stress. In L. Goldberger & S. Breznitz (Eds.), *Handbook of stress: Theoretical and clinical aspects* (pp. 69–87). New York: The Free Press.
32. Ben Zur, H., & Breznitz, S. J. (1981). The effect of time pressure on risky choice behavior. *Acta Psychologica, 47*(2), 89–104.
33. Simon, H. A. (1956). Rational choice and the structure of environment. *Psychological Review, 63*(2), 129–138.
34. Gigerenzer, G., & Selten, G. (2001). *Bounded rationality: The adaptive toolbox*. Cambridge, MA: The MIT Press.
35. Vaughan, D. (1999). The dark side of organizations: Mistake, misconduct, and disaster. *Annual Review of Sociology, 25*, 271–305.
36. Slovic, P., Fischhoff, B., & Lichtenstein, S. (1977). Behavioral decision theory. *Annual Review of Psychology, 28*, 1–39.
37. Peterson, C. R., & Beach, L. R. (1967). Man as an intuitive statistician. *Psychological Bulletin, 68*(1), 29–46.
38. Kahneman, D., Slovic, P., & Tversky, A. (1982). *Judgment under uncertainty: Heuristics and biases*. New York: Cambridge University Press.
39. Tversky, A., & Kahneman, D. (1974). Judgment under uncertainty: Heuristics and biases. *Science, 185*, 1124–1131.
40. Flin, R., Salas, E., Strub, M., & Martin, L. (1997). *Decision making under stress*. Aldershot, England: Ashgate Publishing Ltd.
41. Orasanu, J., & Fischer, U. (1997). Finding decisions in natural environments: The view from the cockpit. In C. Zsambok & G. Klein (Eds.), *Naturalistic decision making* (pp. 343–358). Mahwah, NJ: Erlbaum.
42. Janis, I. L., & Mann, L. (1977). *Decision making: A psychological analysis of conflict, choice, and commitment*. New York, NY: Free Press.
43. Hammond, K. R., & Adelman, L. (1976). Science, values, and human judgment. *Science, 194*, 389–396.
44. Slovic, P., Kunreuther, H., & White, G. F. (1974). Decision processes, rationality, and adjustments to natural hazards. In G. F. White (Ed.), *Natural hazards* (pp. 187–205). New York: Oxford University Press.
45. Bryan, J. L. (2002). Behavioral response to fire and smoke. In P. J. DiNenno (Ed.), *The SFPE handbook of fire protection engineering* (3rd ed., pp. 3-315–3-341). Quincy, MA: National Fire Protection Association.
46. Proulx, G. (2002). Movement of people: The evacuation timing. In P. J. DiNenno (Ed.), *The SFPE handbook of fire protection engineering* (3rd ed., pp. 3-342–3-365). Quincy, MA: National Fire Protection Association.
47. Tong, D., & Canter, D. (1985). The decision to evacuate: A study of the motivations which contribute to evacuation in the event of fire. *Fire Safety Journal, 9*(3), 257–265.
48. Averill, J. D., Mileti, D. S., Peacock, R. D., Kuligowski, E. D., Groner, N., Proulx, G., et al. (2005). *Federal building and fire safety investigation of the World Trade Center disaster: Occupant behavior, egress, and emergency communications* (Rep. NCSTAR 1-7). Gaithersburg,

MD: National Institute of Standards and Technology. Retrieved from http://wtc.nist.gov/NISTNCSTAR1-7.pdf
49. Isner, M. S., & Klem, T. J. (1993). *Fire investigation report on World Trade Center explosion and fire, New York, NY, February 26, 1993*. Quincy, MA: National Fire Protection Association.
50. Bryan, J. L. (1982). Human behavior in the MGM Grand Hotel fire. *Fire Journal, 76*, 37–48.
51. Best, R. L. (1977). *Reconstruction of a tragedy: The Beverly Hills Supper Club fire*. Quincy, MA: National Fire Protection Association.
52. Kuligowski, E. D. (2011). *Terror defeated: Occupant sensemaking, decision-making and protective action in The 2001 World Trade Center disaster*. PhD Thesis, University of Colorado, Boulder.
53. Fahy, R. F., & Proulx, G. (1997). Human behaviour in the World Trade Center evacuation. In Y. Hasemi (Ed.), *Proceedings of the Fifth International Symposium on Fire Safety Science* (pp. 713–724). London, England: Interscience Communications Ltd.
54. Mileti, D. S., & O'Brien, P. W. (1992). Warnings during disasters: Normalizing communicated risk. *Social Problems, 39*(1), 40–57.
55. Proulx, G., & Sime, J. D. (1991). To prevent 'panic' in an underground emergency: Why not tell people the truth? *Fire Safety Science, 3*, 843–852. doi:10.3801/IAFSS.FSS.3-843.
56. Latane, B., & Darley, J. M. (1970). *The unresponsive bystander: Why doesn't he help?* New York: Appleton-Century Crofts.
57. Proulx, G., Latour, J. C., & MacLaurin, J. W. (1994). *Housing evacuation of mixed abilities occupants* (Internal Rep. IRC-IR-661). Institute for Research in Construction, National Research Council of Canada.
58. Bruck, D. (2001). The who, what, where and why of waking to fire alarms: A review. *Fire Safety Journal, 36*, 623–639.
59. *Americans with Disabilities Act*. (1990). U.S. Code Chapter 126.
60. Helweg-Larsen M, Shepperd JA (2001) Do Moderators of the Optimistic Bias Affect Personal or Target Risk Estimates? A Review of the Literature, Personality and Social Psychology Review, Vol. 5, No. 1, 74–95.
61. Sime, J. (1984). *Escape behaviour in fire: 'Panic' or affiliation?* PhD Thesis, Department of Psychology, University Of Surrey, Guildford.
62. Canter, D. (1980). *Fires and human behaviour*. New York: John Wiley & Sons.
63. Wood, P. G. (1972). *Fire Research Note 953*. Borehamwood, England: Building Research Establishment.
64. Quarantelli, E. L., & Dynes, R. R. (1972). When disaster strikes (it isn't much like what you've heard and read about). *Psychology Today, 5*, 67–70.
65. Fischer, H. W., III. (1998). *Response to disaster: Fact versus fiction and its perpetuation: The sociology of disaster* (2nd ed.). New York: University Press of America.
66. Santos, G., & Aguirre, B. E. (2004). Critical review of emergency evacuation simulation models. In R. D. Peacock & E. D. Kuligowski (Eds.), *Workshop on building occupant movement during fire emergencies* (pp. 27–52). Gaithersburg, MD: National Institute of Standards and Technology.
67. Gwynne, S. M. V. (2010). *Conventions in the collection and use of human performance data*. (Tech. Rep. NIST GCR 10-928). Gaithersburg, MD: National Institute of Standards and Technology.
68. Gwynne, S. M. V. (2012). *Translating behavioral theory of human response into modeling practice*. (Grant Rep. NISTGCR—12-972). Gaithersburg, MD: National Institute of Standards and Technology.
69. Kinateder, M., Kuligowski, E. D., Reneke, P. A., & Peacock, R. D. (2014). *Review of risk perception in building fire evacuation* (NIST TN—1840). Gaithersburg, MD: National Institute of Standards and Technology.
70. McConnell, N. C., Boyce, K. E., Shields, J., Galea, E. R., Day, R. C., & Hulse, L. M. (2010). The UK 9/11 evacuation study: Analysis of survivors' recognition and response phase in WTC1. *Fire Safety Journal, 45*(1), 21–34.

71. Gwynne, S., Galea, E. R., Owen, M., Lawrence, P. J., & Filippidis, L. (1999). A review of the methodologies used in evacuation modelling. *Fire and Materials, 23*(6), 383–389.
72. Kuligowski, E. D., Peacock, R. D., & Hoskins, B. L. (2010). *A review of building evacuation models* (2nd ed.) (Tech. Note 1680). Gaithersburg, MD: National Institute of Standards and Technology.
73. Breaux, J., Canter, D., & Sime, J. D. (1976). *Psychological aspects of behaviour of people in fire situations.* In 5th International Fire Protection Seminar (pp. 39–50). Karlsruhe, West Germany, Vereinigungzur Forderung des Deutschen Brandschutzes e.v., 1976.
74. Gwynne, S., Galea, E. R., Parke, J., & Hickson, J. (2003). The Collection and analysis of pre-evacuation times from evacuation trials and their application to evacuation modelling. *Fire Technology, 39*(2), 173–195.
75. Kuligowski, E. D., Peacock, R. D., Reneke, P. A., Wiess, E. A., Averill, J. D., Hagwood, R. C., et al. (2014). *Movement on Stairs during building evacuations*, NIST TN—1839. Gaithersburg, MD: National Institute of Standards and Technology.
76. Reneke, P. A. (2013). *Evacuation decision model.* (NIST Interagency/Internal Rep. NISTIR—79140). Gaithersburg, MD: National Institute of Standards and Technology.
77. Galea, E. R., Shields, J., Canter, D., Boyce, K., Day, R., Hulse, L., et al. (2006). Methodologies employed in the collection, retrieval and storage of human factors information derived from first hand accounts of survivors of the WTC disaster of 11 September 2001. *Journal of Applied Fire Science, 15*(4), 253–276 (published in Nov 2008).

Chapter 4
Verification and Validation Methods

Weiguo Song and Libi Fu

4.1 Introduction

Evacuation modelling has been a useful tool for researchers to investigate pedestrian dynamics. Many evacuation models have been proposed, e.g., the social force model [1], visual hindrance model [2], lattice gas model [3], multi-grid model [4] and floor field model [5]. They are usually used to study pedestrian movement characteristics and crowd behaviours such as lane formation in bi-directional flow, self-organization phenomena in human crowds and herding behaviour in massive crowd evacuation. With the development of computer technology, it enables users to produce sophisticated crowd behaviours and then distinguish individual members of a large population in a model with the graphical user interface.

Users can also simplify an evacuation scenario to obtain values of parameters. Regardless of the level of sophistication of the graphical user interface, evacuation modelling is only a tool employed to aid engineers in studying evacuation dynamics [6]. When establishing evacuation models, an important issue is how to ensure their quality. A computer program with bugs may produce incorrect results or subtle nonsense. A model not been validated may result in wrong decisions [7]. Plausible parameter settings or assumptions of common sense always raise a question of whether those models can accurately and quantitatively describe evacuation scenarios. Thus the process of verification and validation (V&V) is necessary in evacuation modelling.

Verification and validation (V&V) methods are always employed to evaluate the degree of model authenticity and define the scope of model application. Evacuation models are no exception [8]. The verification process addresses the quality of simulation treatment of a model used in the prediction. It is rooted in issues of continuum

W. Song (✉) • L. Fu
SKLFS, University of Science and Technology of China, Hefei, Anhui 230026, P.R. China
e-mail: wgsong@ustc.edu.cn; fulibi@mail.ustc.edu.cn

A. Cuesta et al. (eds.), *Evacuation Modeling Trends*,
DOI 10.1007/978-3-319-20708-7_4

and discrete mathematics and in the accuracy and correctness of complex logical structures (computer codes). The validation process addresses the quality of models. It determines the degree to which a model is an accurate representation of the real world from the perspective of the intended uses of the model. Modelling results within an accepted deviation of reality serve as a precondition to use pedestrian models. Nevertheless, pedestrian behaviours alter according to local conditions such as densities, direction of interaction between pedestrians (walking towards each other, crossing, overtaking) and internal states determined by age, culture and purpose of a trip [9]. Validation methods are needed to identify the adequacy of these models.

To date, there has not been a common standard procedure to execute the process of V&V particularly outlined for building evacuation models. The International Standards Organization just provided a framework for assessment, verification and validation of all types of calculation methods used as tools for fire safety engineering [10]. The International Maritime Organization (IMO) proposed a principal guidance, i.e. the MSC/Circ.1238 (Guidelines for evacuation analysis for new and existing passenger ships) [11], which analysed the V&V of evacuation models. These guidelines are not only applied to maritime evacuation scenarios, but also used to test evacuation models in other areas, e.g., buildings and other means of transportation [12]. Prof. Galea et al. [13] addressed some limitations of these guidelines, and presented two evacuation model validation data sets from trials at sea. They assessed the goodness of fit between predicted model data and experimental data through three objective measures. Klüpfel [14] discussed various aspects of ship evacuation: Guidelines, simulation, validation and acceptance criteria according to the guidance developed by IMO. Ronchi et al. [12] reviewed current procedures, tests and methods to assess the V&V of evacuation models. Issues on the definition of the acceptance criteria of a standard V&V protocol were discussed. They tended to address the variability of evacuation model predictions concerned with the number of runs and application of stochastic algorithms. This provided recommendations to improve the V&V methods described in the MSC/Circ.1238.

In spite of all the previous verification suggestions and validation investigations, there are still some crowd dynamics models that neglect a systematic process of verification and validation. This may be due to several factors: Firstly, pedestrian crowds, flow conditions and geometries vary under different situations; secondly, pedestrian movement, which often needs well controlled laboratory conditions to obtain higher reliability, is difficult to identify due to partial inadequacy of measurement tools. These factors induce rare systematic empirical data. The quality and quantity of data are of significant importance to validate crowd dynamics models beyond the most basic aspects such as mean velocity [15]. Campanella et al. [9] emphasized that researchers usually performed a limited amount of validation assessment, and very few considered the applicability of a model in different walking situations and population characteristics. Zhang et al. [16] presented a simulation model for passenger walking behaviour considering the circumstance and interactions of passengers in transport hubs. They used the fundamental diagram test and bi-directional walking experiment run for the verification and validation.

However, they also stated that more observations had to be conducted, and the parameter calibration and validation should be extended in further research. Schadschneider and Seyfried [17] emphasized that much attention should be paid to quantitative validation, especially regarding the fundamental diagram. Gwynne et al. [6] demonstrated that although some behavioural models had been implemented, no evacuation model completely performed all the identified evacuation behaviours. What is more, not all these evacuation behaviours were fully understood or quantified.

It is a complex task to identify the values of parameters in microscopic pedestrian models. This is partially due to a large number of parameters that are not easy to observe directly, or few suitable data. For instance, when we identify crowd behaviour in emergency conditions in a model, rare real-world data describing the characteristics of emergent behaviour in corresponding situations will always be a bottleneck in specific parameter settings. Even the researchers who proposed the few existing models of emergent crowd behaviour have stated the need for more rigorous modelling frameworks and the development of approaches to evaluate the reliability of model predictions [18]. Therefore, an understanding of the role of different forms of V&V methods in the general acceptability of evacuation models is of paramount importance.

In the following, we discuss a set of recommended verification tests and examples of validation tests. Limitations of V&V methods are analysed. Current literature on V&V approaches employed in evacuation modelling is also reviewed. Finally, conclusions are reported.

4.2 Verification Methods

4.2.1 Definition

Verification can be defined as "The process of determining that a model implementation accurately represents the developer's conceptual description of the model and the solution to the model" [19]. It is also described as "solving the equations right". This shows that verification is related more to mathematics rather than engineering. The objective is debugging the computer program.

The process of verification checks whether conceptual models are correctly implemented in the computational models, and whether computer code is correct. It is intended to examine errors in the model performance and solution, i.e. verifying the code and calculation. The verification of the code is discovering and deleting errors in it, that is, error evaluation. It involves discovering bugs, wrong implementations of conceptual models, incorrect inputs and errors in usage. Verifying a calculation is described as error estimation which identifies the accuracy of a calculation. It is concerned with error bands, grid convergence study [20, 21] and so on. It is noteworthy that verification should not be conducted with empirical data.

4.2.2 *Verification Tests*

4.2.2.1 General Verification Methods

Once a model has been programmed, programmers have to check if there are some bugs or errors in computer code. In general, some techniques are available in this process. They mainly include: (1) good programming practice, (2) verifying intermediate simulation outputs, (3) final simulation outputs in contrast to analytical results and (4) animation [7].

1. **Good programming practice**
   Soft engineering involves a large research area. There have been a large number of procedures for creating great computer programs and verifying corresponding software, such as modular programming, object oriented programming, chief programmer's approach, structured walk-throughs and correctness proofs. Models usually consist of many program codes. A sound suggestion is modular programming. Hence a complex program is divided to be easily and clearly verified. Good programming practice is one of the most fundamental methods to ensure a successful computer program.
2. **Verifying intermediate simulation outputs**
   There are two methods to check intermediate simulation outputs, namely manual calculations compared with these simulation outputs and automatically obtaining results from a computer program. The latter is also described as tracing. If programmers dislike manually calculating intermediate simulation outputs, they can also 'eyeball' the trace of a program, and discover code errors. Advanced tracing facilities and debuggers are becoming popular in modern simulation software.

   As demonstrated above, good programming practice always needs designers to divide a program, and then check it modular by modular. This process may be finished by different members of a team. Some programmers may verify the pseudorandom number generator since it produces numbers that are not statistically independent and uniformly distributed in practice. Altering a new generator may induce a better result. In addition, programming errors may occur in the generator which requires a higher language to modify. The programmers may further test subprograms which produce samples from e.g. non-uniform distributions. Some software may use the notation $N(\mu, \sigma)$ as the normal distribution instead of $N(\mu, \sigma^2)$. This will confuse users. Other confusions such as the parameter λ in exponential distributions (λ denoting the average interarrival time rather than the arrival rate) and wrong units may also arise in programs.

   In order to identify that subprograms generate effective random variables, programmers should first read these subprograms. Then they can assess the mean and variance of samples, and contrast them with theoretical values. Statistic tests such as the *t* test and chi-square test can be employed to detect deviations between observed statistics and theoretical values.

3. **Final simulation outputs in contrast to analytical results**
   A queuing simulation program may produce final outputs after processing a large number of customers. It is actually impossible for programmers to verify this type of simulation responses through manual calculations or tracing. In this case, programmers may conduct a simplified simulation program with a recognized analytical solution. Of course these programmers should be instructed by knowledge of theoretical models with noted solutions when they address real systems. A large quantity of literature or books on this topic can provide guides. Statistical techniques are also useful in the verification of simulation outputs.
4. **Animation**
   Animation is significant for programmers to display a dynamic system. It can be used to discover simulation errors or bugs in contrast to corresponding real system. Due to its intuitive interface, animation is welcome especially in conducting evacuation modelling or traffic models. However, it is noteworthy that animation may induce programmers to concentrate on short simulation processes, and thus neglect long runs. This is dangerous. Experienced programmers may seek to create long runs, and then display them to users.

4.2.2.2 Verification of Evacuation Models

Model verification and validation include component and functional testing, and qualitative and quantitative validations. Component and functional testing belongs to the process of verification. Component testing is performed to verify that each component of a model is effective, and the produced results are satisfactory. Functional testing examines whether the relevant results of this model reach the modelling objectives. When establishing an evacuation model, the first and most important process is to examine whether its individual components are functioning correctly. Meanwhile, functional testing should be conducted, and the program can be modified to produce relevant results [8].

Component testing is a code fragment testing process at the lowest level. For example, users can test random number generators or compare output results of the same program written in different platforms. Moreover, computer code can be divided into segments, and then be verified individually [22].

At the highest level, users can perform some test scenarios to verify model subcomponents. In terms of a walker model, a common verification method is to demonstrate that this model is able to reproduce some basic characteristics of pedestrian movement. For example, pedestrians should not be trapped in corners, but be able to avoid collisions. Moreover, the model may be capable of generating self-organized phenomena such as the lane formation or reproducing reasonable fundamental diagram relations (e.g., speed decreasing with an increase in density) [9]. Further component testing such as geometrical tests, door width testing, barrier models, speed/density, histogram inputs and speed/density measurements as outputs can be addressed.

Functional testing is a specific verification process. A series of geometric models such as corridors, corners, unidirectional and bidirectional flow tests can be employed to check whether evacuation modelling produces expected results. The speed/density test or 100 m dash test (i.e. a pedestrian, whose movement speed is 1 m/s, walks 100 m with 100 s) can also be used to examine whether a model is working satisfactorily.

Helbing and Molnár [23] proposed the social force model which simulated pedestrian dynamics in different situations. They verified this model through describing the self-organization of collective phenomena of pedestrian behaviour and oscillatory changes of the walking direction at narrow passages. The floor field model presented by Kirchner et al. [5] was also verified through qualitative assessments. This model reproduced different types of pedestrian behaviour during the evacuation from a large room with one or two doors.

Teo [8] introduced an evacuation model, namely EvacuationNZ. They conducted verification testing to ensure that some model components produced expected results. Two components, i.e. door queuing and movement on stairs, were involved. Here, we briefly summarize their verification process as follows:

Verify 1. Door Queuing

This test analysed different correlations of flow rates through doors, and then compared manual calculations with model outputs.

Scenarios

A simple two-node map with a door linking these two nodes was considered. One thousand evacuees were in the starting node. The door width altered. Accordingly, the flow rate was calculated. Two initial values of crowd density in the starting node were analysed. The first was the maximum crowd density of 2.0 persons/m^2, and the second was a low crowd density of 0.5 persons/m^2 which enabled pedestrians to move at a maximum potential speed. Thus the required node areas were 500 m^2 and 2000 m^2.

Expected Results

The evacuation time and flow rate of the two crowd density scenarios were consistent with the expected manual calculations.

Test Method

This test method was a quantitative verification of model results. The difference between simulated results and expected results was discussed.

User Actions

Users could improve this test by setting different pedestrian movement speeds. Multiple runs of a scenario should be conducted so that results were effective.

Verify 2. Movement on Stairs

This test verified occupant movement on stairs. Simulation results were compared with manual calculations.

Scenarios

Two nodes representing the stairs and safe area were linked by the stair connection. Two values of occupant density of starting nodes, namely 2.0 persons/m^2 and 0.5 persons/m^2, were employed. The stair dimensions were changed. With regard to a staircase with 1.0 m wide and 100 m long, there were 200 and 50 occupants for high and low crowd density nodes, respectively.

Expected Results

The actual evacuation time and flow rate were similar to manually calculated results.

Test Method

This test was a process of quantitative verification. Simulation results were compared with manual calculations.

User Actions

It was recommended that the random initial pedestrian distribution and movement speed should be discussed to analyse their influence on evacuation time and flow rate. A test should be repeated several times.

Song et al. [24] verified their proposed model (CAFE) by comparison with the social force model. Evacuation time, arching, clogging, faster-is-slower, door width etc. were involved.

Hostikka et al. [25] presented the verification process of a human movement algorithm in FDS5+Evac according to the IMO document. They performed 11 test cases where some simple geometries and smoke effect were employed to detect pedestrian movement speed. The input files and simulation results were displayed on the FDS+Evac web page (http://www.vtt.fi/proj/fdsevac/?lang=en).

Fang et al. [26] considered the effects of fire products on evacuation in a multigrid model, and verified evacuation time of this model against FDS+Evac. Capote et al. [27] compared their proposed model, namely EvacTunnel, with other evacuation models, e.g., GridFlow, STEPS and Pathfinder. Two verification tests were conducted. Test 1 performed no pedestrian behaviour to check pedestrian movement. Test 2 compared evacuation time of the proposed model with others.

Ronchi et al. [12] elaborately presented five principal components of evacuation models, i.e. pre-evacuation time, movement and navigation, exit usage, route availability and flow conditions. These components were the most basic elements of an evacuation model for presenting a scenario. They analysed two categories of tests, namely analytical verification (AN_VERIF) and the verification of emergent behaviours (EB_VERIF), partially according to the MSC/Circ.1238 proposed by the IMO [28]. Analytical verification represented component testing that derived results through mathematical formulae or evidence. The verification of emergent behaviours checked if evacuation models qualitatively generated results that reflected real crowd behaviours in emergency situations. Table 4.1 presents some suggested verification tests given core components and sub-elements of evacuation models, as discussed in Ref. [12]. It provides clear reference for verification tests.

The suggested test methods involve a quantitative and qualitative evaluation of model results. The former is concerned with differences between expected results and simulation results. The latter is according to qualitative observations of simulation results or visualization outputs and a comparison between expected behaviour and the contemporary behavioural theory. Though we can use these test methods

Table 4.1 Suggested verification tests for evacuation modelling

Core component	Sub-element	Suggested tests	Type of test
Pre-evacuation time	Pre-evacuation time distributions	Modified IMO Test 5	AN_VERIF
Movement and navigation	Speed in a corridor	IMO Test 1	AN_VERIF
	Speed on Stairs	IMO Test 2 and IMO Test 3 (if necessary)	AN_VERIF
	Movement around a corner	IMO Test 6	AN_VERIF
	Assigned demographics	Modified IMO Test 7	AN_VERIF
	Reduced visibility vs. walking speed	New test	AN_VERIF
	Occupant incapacitation	New test	AN_VERIF
	Elevator usage	New test	AN_VERIF
	Horizontal counter-flows (rooms)	Modified IMO Test 8	EB_VERIF
	Group behaviours	New test	EB_VERIF
	People with movement disabilities	New test	EB_VERIF
Exit usage	Exit route allocation	Modified IMO Test 10	AN_VERIF
	Social influence	New test	EB_VERIF
	Affiliation	New test	EB_VERIF
Route availability	Dynamic availability of exit	New test	AN_VERIF
Flow conditions	Congestion	Modified IMO Test 11	EB_VERIF
	Maximum flow rates	IMO Test 4	EB_VERIF

from Ref. [12], specific models are recommended to examine their assumptions to ensure the accuracy of usage.

4.2.3 Limitations

As described above, researchers have already performed many verification tests to examine modelling results. However, most of program components are tested individually. It is suggested that different combinations of components should also be verified to determine whether simulation results are satisfactory.

Functional verification identifying safety limits in an environment is an ongoing process. For example, an evacuation model employed to check its relevant results compared with guidelines should be used to calculate specific variables such as maximum travel distances and available exit widths according to the building code. In addition, different conditions require corresponding building codes to evaluate modelling results [22].

Both component and functional testing are prior to the validation process. Successfully addressing these verification cases does not imply that an evacuation model is validated or able to predict real evacuation performance. In order to achieve convincing and intended evacuation modelling, validation tests are necessary.

4.3 Validation Methods

4.3.1 Definition

Validation is defined as "The process of determining the degree to which a model is an accurate representation of the real world from the perspective of the intended use of the mode". It is also described as "solving the right equations" [19].

The process of validation is to check whether computational simulation is consistent with real world observations. Experimental data are employed to identify errors and uncertainty of simulation results. Nevertheless, there may be bias or random errors of experimental data. Therefore, the experimental data set contains suitably quantifying and documenting these errors. Various levels of accuracy should be available in the validation activities according to application [19].

4.3.2 Validation Tests

Validation testing is always divided into two categories, namely qualitative and quantitative validation. A qualitative test is to determine whether model predictions are plausible on an aggregate (macroscopic) level. Qualitative validation can be conducted to contrast the nature of predicted human behaviour with the educated expectations of human behaviour in a certain situation. This validation is subjective to some degree but is a need to ensure that a program can reproduce realistic human behaviour [8].

Quantitative validation compares model predictions with reliable values which are always from real experimental data. It is also available to adopt results from other evacuation models or calculation methods. A typical validation will analyse the relationship between empirically obtained data concerned with pedestrians and evacuation model outputs. Ideally the assessment is performed reproducing similar crowd characteristics, walking configurations and pedestrian flow [9].

Validation methods in current evacuation models include [29]: validation against code requirements, validation against fire drills or other pedestrian movement experiments/trials, validation against literature on previous evacuation experiments (flow rates, etc.), validation against other models and third party validation. Here, we principally describe validation against experimental data which include empirical data and controlled experimental data. The metric employed to objectively assess evacuation modelling results is also presented.

4.3.2.1 Validation Against Empirical Data

Empirical data include observational data and fire drill data. Observational data are primarily concerned with analysing video recordings of pedestrian traffic or emergency situations such as earthquake. A common characteristic of these data is that they are affected by boundary and initial conditions which are difficult to be accurately quantified. Thus empirical results from complicated situations are more suitable in qualitative validations, e.g. lane formation in counterflow, oscillations in bottlenecks, speed decreasing with an increase in crowd density and merging behaviour in the floor–stair interface. Ma et al. [30] proposed a metric distance based model and a K-Nearest-Neighbor (KNN) counterflow model. The self-organized lane formation was captured, and factors affecting the number of lanes formed in a channel were investigated. The lane formation pattern of the KNN counterflow model was compared with real pedestrian traffic while crossing a street. An ultra-high rise building evacuation model which simulated both pedestrian movement and elevator transportation was presented by Ma et al. [31]. The mass evacuation process focused on the effect of merging behaviour was designed to validate this model. After validation and calibration, the proposed Ultra high-rise building Evacuation with Elevators (UEE) model was adopted to explore two main evacuation strategies, i.e., phased evacuation and total evacuation with the aid of elevators.

The influence of different structures of the floor–stair interface on pedestrian merging behaviour through modelling and comparing with empirical data was also investigated by Huo et al. [32]. Xu and Song [33] studied an evacuation process of a four-story building by means of an egress drill, and then reproduced it through the multi-grid modelling.

Quantitative validation against literature on previous experimental results, especially fundamental diagrams, is a common way. Fundamental diagrams studied by Predtechenskii and Milinskiĭ [34], Weidmann [35] or Older [36] have been widely used to quantitatively validate evacuation modelling. Ronchi et al. [12] discussed a set of potential data-sets for the analysis of core behavioural components of evacuation models. Detailed information could be obtained in the corresponding reference. However, this was not a complete list of validation tests. Liu et al. [37] proposed an agent-based microscopic pedestrian flow simulation model, namely CityFlow, and validated it through comparing simulation results of the density-speed fundamental diagrams with empirical data. Campanella et al. [9] presented quantitative and qualitative validation using different parameter sets and types of flow. They quantitatively assessed travel time, fundamental diagrams and bottleneck capacity compared with experimental data. The quantitative assessment results were concluded by using simple validation scores.

Zhou et al. [38] proposed a pedestrian evacuation time model for Urban Metro Hubs (UMHs). The evacuation characteristics, speed-density variation law, pedestrian drop-of time, platform evacuation time and channel evacuation time were validated through analysing video sequences of pedestrian flow in a metro hub. With regard to pedestrian flow experimental data, a large number of researchers [39–46] have conducted field study, and have obtained huge amounts of experimental data, which are available for validation of pedestrian movement characteristics. Figures 4.1a, b are empirical fundamental diagrams from different scenarios investigated by researchers [2, 34, 35, 47–49]. Different types of experimental data on fundamental diagrams, bottlenecks and doors can also be obtained on the web (http://www.ped-net.org/index.php?id=12&ID=48). These empirical data are always used to validate evacuation modelling.

4.3.2.2 Validation Through Controlled Experiments

As stated above, due to the uncertainties of observational data or fire drills, conducting controlled experiments tends to be a favourable method for validation. Thus the problem of boundary and initial conditions is solved. A large number of researchers have designed controlled experiments to examine pedestrian movement features and validate related evacuation models.

Campanella et al. [9] qualitatively analysed bidirectional flows, unidirectional flows and narrow corridor bottleneck flows of simulation results validated through controlled experiments. Avoidance behaviour and characteristics of different types of flows, e.g., lane formation in bidirectional flows, the zipper-effect in a narrow corridor and funnel shape in front of a bottleneck, were discussed. We briefly summarize their qualitative assessments in Table 4.2 where validation criteria and assessments of models were proposed according to related experiments. Fang et al. [51] built a continuous distance model (CDM) for the single-file pedestrian movement. The continuous step length and dependency between the transition probability and distance headway were adopted in the movement algorithm. Simulation

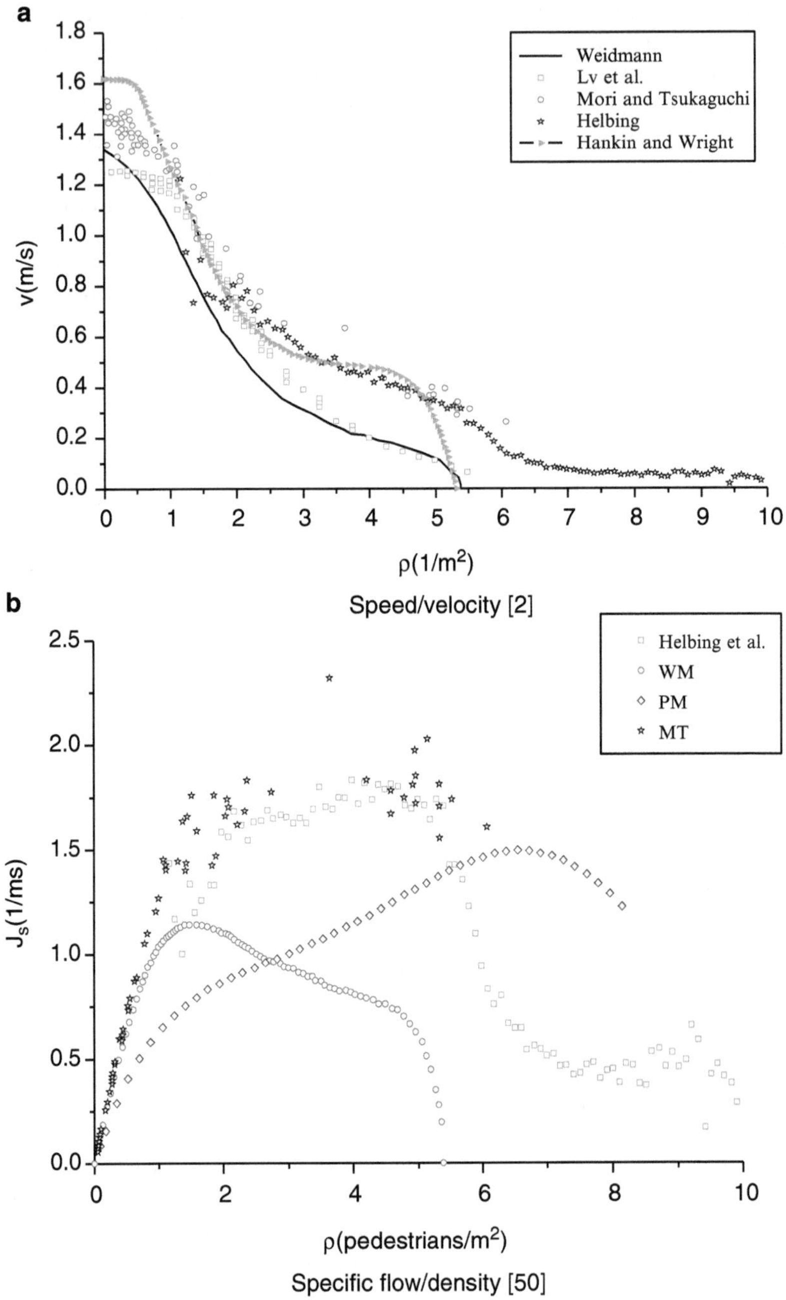

Fig. 4.1 Fundamental diagrams of pedestrians from different regions of the world. (**a**) speed/velocity [2]; (**b**) specific flow/density [50]

Table 4.2 Qualitative assessments of three walking situations

Walking situations	Validation criteria	Assessments		
		Good	Medium	Bad
Bidirectional flows (no frontal and lateral collisions between pedestrians during the experiments, no solid wall in the walking area, pedestrians always staying inside the delimited area)	The formation of self-organized lanes and their states	Lanes are formed and their states are dynamically stable	Lanes are formed but their states are unstable	Lanes are not formed
	Avoidance behavior (a): How pedestrians behave when walking straight on	Never collide	Few collisions	Many collisions
	Avoidance behavior (b): How often they collide with nearby pedestrians	Never collide	Few collisions	Many collisions
	Avoidance behavior (c): Are pedestrians being pushed backwards in straight on collisions?	Never pushed	Few pedestrians	Many pedestrians
	Staying in the simulated area	Always staying	Few pedestrians walking outside	Many pedestrians walking outside
Unidirectional flows (no frontal and lateral collisions between pedestrians during the experiments, no solid wall in the walking area, pedestrians always staying inside the delimited area)	The avoidance behavior (a): Average distances that pedestrians keep from each other	Similar distances	Larger distances	Very large distances
	The avoidance behavior (b): How often they collide with nearby pedestrians	Never collide	Few collisions	Many collisions
	Staying in the simulated area	Always staying	Few pedestrians walking outside	Many pedestrians walking outside

(continued)

Table 4.2 (continued)

Walking situations	Validation criteria	Assessments		
		Good	Medium	Bad
Narrow corridor bottleneck flows (congested flow up and downstream of a bottleneck)	The formation of the self-organized zipper-effect inside the corridor	Pedestrians are displaced laterally and stay stable at this position	Positions are not permanent	Pedestrians follow each other
	A funnel shape is formed upstream of the bottleneck in congested flow	Funnel is formed	Funnel is not stable	Never formed
	The avoidance behavior (a): How often they collide with nearby pedestrians	Never collide	Few collisions	Many collisions
	The avoidance behavior (b): Slow pedestrians are trapped outside the corridor	Never be trapped	Few be trapped	Many be trapped
	The avoidance behavior (c): Are pedestrians being pushed towards obstacles?	Never pushed	Few pedestrians	Many pedestrians
	Staying in the simulated area	Always staying	Few pedestrians walking outside	Many pedestrians walking outside

results of the flow-density and velocity–density were compared with experimental results. Lv et al. [2] developed a 2-D continuous model that integrated a self-slowing, local direction-changing mechanism and visual hindrance information. They conducted three experimental scenarios, i.e., passage, bottleneck and class-room evacuation, to validate this model. Stop-and-go waves and lane formation were reproduced. The fundamental diagram and specific flow obtained from simulation results agreed well with experimental data.

In addition to these simple scenarios stated above, evacuation trials in complex scenarios are also performed to validate evacuation modelling. In the case of evacuation, it is rather strictly limited in space and time, i.e. its beginning and end are well-defined (sound of the alarm, initial position of all individuals, safe areas, final position of all individuals etc.), since the aims and routes are known and usually the same, e.g. the exits and egress routes [52]. Evacuation trials range from public buildings, industrial premises to passenger vessels [53]. From these experiments, evacuation time, pre-evacuation time and pedestrian movement speed can be achieved.

Helbing et al. [54] employed the lattice gas model to simulate an evacuation process from a classroom. Simulation results of flow rate were compared with experimental data. They agreed well with each other. Zhang et al. [55] developed the multi-grid model to analyse the influence of pre-movement time on the evacuation process of students from a classroom, and simulation results were consistent with experimental data. Fang et al. [56] reproduced an evacuation process in a teaching building with two neighboring exits by a modified biased random walk model. Simulation results of the cumulative distribution of the instantaneous velocity and fundamental diagram of the flow through exits agreed well with experimental data.

Hostikka et al. [25] validated FDS+Evac by performing an evacuation experiment of a public library. This was to study the capability to predict the entire movement phase of the evacuation, consisting of movement inside a floor, queuing to a staircase and finally movement through a narrow staircase to an exit.

Galea et al. [13] validated the ship evacuation software, i.e. maritimeEXODUS, with data sets of passenger response times, starting locations etc. generated from ship evacuation.

Requiring the high number of repetitions is always a problem in performing evacuation trials. The well-being of participants is one major concern in evacuation exercises. Because of practical, financial and ethical constraints, an evacuation trial cannot be realistic by its very nature [52]. Therefore, an evacuation exercise does not convey the increased stress of a real evacuation. Thus deviations between simulation results and experimental data exist and should be analysed.

Numerical simulations of escape panic of confined pedestrians have revealed some interesting dynamical characteristics such as pedestrian arch formation around an exit, self-organized queuing and scale-free behaviour [57]. As data under emergency and panic conditions are difficult to capture, they are rare. Attention must be paid to the integrity of data, suitability of an experiment and repeatability of the experiment. In practice, there are issues concerned with the nature of crowd safety that can never fully be tested [22]. Researchers have sought to validate models' predictions through experiments with biological agents such as ants and mice. These experiments have identified that ants naturally formed collective traffic and followed physical pathways which were closely related to crowd movement. Moreover, ants are easy to handle, and experimental equipment is simply [53]. These advantages result in further research on examining crowd behaviour and validating evacuation models with experiments on biology agents.

Saloma et al. [57] investigated escape panic of confined pedestrians by conducting experiments of mice escaping out of a water pool and comparing escape behaviour of mice with panicked humans predicted from computer models.

Dussutour et al. [58] presented an experimental study of ants confronted with two alternative routes. The mechanism of 'symmetry-restoring transition' was identified by analytical calculations and Monte Carlo simulations.

Altshuler et al. [59] demonstrated the existence of a nonsymmetrical use of two exits in panic conditions by conducting ant experiments. Then they reproduced the observed escape dynamics through modifying a previous theoretical model.

Ballerini et al. [60] showed that interaction among individuals did not depend on the metric distance, but rather on the topological distance. This result was obtained through analysing bird flocking events from a field study and numerical simulations.

Shiwakoti et al. [61] proposed a mathematical model for crowd panic, and this model was validated by experiments with panicking Argentine ants. They also compared traffic dynamics of humans and Argentine ants to establish a model that simulated pedestrian movement [62]. Through measuring some parameter values on ants and some scaled from human values to ant values according to body mass difference, the model correctly predicted two important properties of crowd behaviour, namely the flow rate and distribution of time headways between successive ants in an escape sequence.

Dias et al. [63] utilized data collected through experiments with panicking ants to investigate the influence of complex configurations such as turning, crossing and merging on the collective egress of crowds. Their empirical study was useful in validation of evacuation modelling related to emergency and high density conditions.

Validation through biological experiments is a new trend. Controlled experiments with biology agents can improve the reliability of evacuation model predictions. However, they cannot be completely relied upon when scaled up to a real human situation [53]. Thus validation through this method is a challenging and ongoing research process.

4.3.2.3 Validation Metric

In validation testing, it is particularly important to quantify the level of agreement between predicted and measured performance. Objective measures to determine the suitability of an evacuation modelling tool are necessary. The metric used to quantify the level of agreement between predicted and measured values comprises three measures [13], as shown in (4.1)–(4.3).

$$\frac{\|E-m\|}{\|E\|}=\frac{\sqrt{\sum_{i=1}^{n}(E_i-m_i)^2}}{\sqrt{\sum_{i=1}^{n}E_i^2}} \tag{4.1}$$

$$\frac{\langle E,m\rangle}{\|m\|^2}=\frac{\sum_{i=1}^{n}E_i m_i}{\sum_{i=1}^{n}m_i^2} \tag{4.2}$$

$$\frac{\langle E,m\rangle}{\|E\|\|m\|}=\frac{\sum_{i=s+1}^{n}\frac{(E_i-E_{i-s})(m_i-m_{i-s})}{s^2(t_i-t_{i-1})}}{\sqrt{\sum_{i=s+1}^{n}\frac{(E_i-E_{i-s})^2}{s^2(t_i-t_{i-1})}\sum_{i=s+1}^{n}\frac{(m_i-m_{i-s})^2}{s^2(t_i-t_{i-1})}}} \tag{4.3}$$

The n-dimensional vector $E=(E_1, E_2, \ldots E_n)$ represents the series of measured experimental data, and E_i denotes the measured value for the *ith* individual. Similarly, the vector $m=(m_1, m_2, \ldots m_n)$ represents the series of predicted model data, and m_i denotes the predicted value for the *ith* individual. t in (4.3) denotes the spacing of data. s in (4.3) represents the period of noise in data or variations in experimental data, because of microscopic behavior that is difficult to be reproduced in a model. A value of s that is greater than the period of the noise in the data provides a method to smooth out the effect of the noise. Nevertheless, a too large vale of s may render the natural data variation lost, and a too small vale of s may induce the variation in the data created by noise to dominate the analysis. Therefore, a suitable value of s depends on the number of data points given by n, and a low value of the ratio s/n is optimal.

Equation (4.1) defines the Euclidean Relative Difference (ERD), which is employed to evaluate the average difference between experimental data (E_i) and model data (m_i). If two curves are identical in magnitude, (4.1) should return a value of 0. A smaller value of (4.1) will produce a better agreement. Equation (4.2) defines the Euclidean Projection Coefficient (EPC). In this formula, the distance between model (m) and experimental (E) vectors will be reduced to its minimum when each experimental data point (E_i) is multiplied by each model data point (m_i). Hence, a measure of the best possible level of agreement between the model (m) and experimental (E) curves is provided by the EPC. An EPC of 1.0 demonstrates that the difference between the model (m) and experimental (E) vectors is as small as possible. Equation (4.3) defines the Secant Cosine (SC). Different from the other two measures, this equation provides a measure of how well the shape of the model data curve matches that of the experimental data curve. It uses the secants, which approximate to tangents, through both curves. An SC of 1.0 demonstrates that the shape of the model (m) curve is consistent with that of the experimental (E) curve. If ERD=0.0, it need not involve the other two measures as experimental data and model data are identical. In all other cases, we had better consider these three measures together in order to identify a good indication of the level of agreement between model and experimental data. In general, a perfect match is to ensure that ERD=0.0, EPC=1.0 and SC=1.0.

4.3.3 *Limitations*

A limited amount of convincing validation assessment and comparisons is observed in many models. For the most part this is due to a general lack of data suitable for validation purposes. The variability of human behavior induces repeatability of experiments to be an issue. It is thus important that different forms of validation (e.g., qualitative and quantitative) have to play the role in the general acceptability of these models. Until a systematic and graduated approach to validation is adopted by authorities, this will be one of the most important problems impeding both the development and wide scale acceptance of evacuation models [6].

A comprehensive list of validation tests should involve experimental data relating to a full range of possible evacuation behaviors and scenarios. However, a limited number of experimental data-sets render the validation of all aspects of evacuation modeling tools difficult. In addition, evacuation models may be conducted starting from a set of specific experimental data. The validation of a model should not be performed with only the same data used for its development. This may generate a circular logic that limits the extent to which model predictions may be generalized for all possible scenarios [12].

With respect to the Guidelines proposed by the International Maritime Organization (IMO), although it is sufficiently developed to address realistic simulations of evacuation onboard, validation data and practical experience on its application are still not enough. Almost all the data and parameters given in the Guidelines depend on well-documented data from civil building experience. Ongoing research and development is necessary to improve the Interim Guidelines [28].

Experiments with biological agents (e.g., ants) are available to investigate crowd behavior under panic or emergency conditions and validate evacuation modeling. Employing biological agents as an alternative empirical system to validate models is an ongoing research process. In future, more algorithms, design solutions and systems modeled for pedestrian traffic with contribution from social insects and other social animals are possible [53].

4.4 Summary

In this chapter, we present verification and validation methods of evacuation modeling. Verification tests and validation tests are discussed. In evacuation modeling, verification is to ensure that computer code of a model is right programmed, or calculations are correct. Then values of related parameters need to be calibrated according to empirical data or controlled experiments. These calibrated parameters are used to conduct simulations of pedestrian dynamics in different scenarios. Simulation results are validated against experimental data.

General verification methods include good programming practice, verifying intermediate simulation results, final simulation outputs compared with analytical results and performing animation. With regard to evacuation modeling verification, component and functional testing are presented. We review some researchers' study on verification methods of evacuation modeling, and briefly summarize their verification processes. Validation serving as a further assessment of an evacuation model is clearly described. Qualitative and quantitative validation tests are involved. Though many validation methods can be used to evaluate simulation results of a model, validation against empirical data and controlled experimental data is principally presented in this chapter. Empirical data are more suitable to qualitative validate evacuation modeling since they are difficult to accurately define initial and boundary conditions, but are sometimes employed to quantitatively assess modeling results, e.g. the fundamental diagrams proposed by some well-known researchers. Validation through controlled experiments is widely used to examine simulation results as initial and boundary conditions are easily defined. Controlled experiments range from simple scenarios such as bidirectional flows, unidirectional flows and narrow corridor bottleneck flows to complex scenarios such as evacuation trials in buildings or passenger vessels. Moreover, biological experiments are also employed to validate evacuation modeling under emergency situations. This is a new trend, and ongoing research is needed. The validation metric which is used to objectively determine the suitability of an evacuation modelling tool is also described.

Though many V&V methods of evacuation modeling have been proposed, there is not a complete verification and validation frame to assess evacuation model results. This is due to behavioral uncertainty in evacuation modeling, a lack of repeated experimental data of single evacuation scenarios etc. Further research on V&V methods of evacuation modeling is needed.

References

1. Helbing, D., Farkas, I., & Vicsek, T. (2000). Simulating dynamical features of escape panic. *Nature, 407*, 487–490.
2. Lv, W., Song, W. G., Ma, J., & Fang, Z. M. (2013). A two-dimensional optimal velocity model for unidirectional pedestrian flow based on pedestrian's visual hindrance field. *IEEE Transactions on Intelligent Transportation Systems, 14*, 1753–1763.
3. Muramatsu, H., & Nagatani, T. (2000). Jamming transition in two-dimensional pedestrian traffic. *Physica A: Statistical Mechanics and its Applications, 275*, 281–291.
4. Song, W., Xu, X., Wang, B.-H., & Ni, S. (2006). Simulation of evacuation processes using a multi-grid model for pedestrian dynamics. *Physica A: Statistical Mechanics and its Applications, 363*, 492–500.
5. Kirchner, A., & Schadschneider, A. (2002). Simulation of evacuation processes using a bionics-inspired cellular automaton model for pedestrian dynamics. *Physica A: Statistical Mechanics and its Applications, 312*, 260–276.
6. Gwynne, S., Galea, E. R., Owen, M., Lawrence, P. J., & Filippidis, L. (1999). A review of the methodologies used in evacuation modelling. *Fire and Materials, 23*, 383–388.
7. Kleijnen, J. P. (1995). Verification and validation of simulation models. *European Journal of Operational Research, 82*, 145–162.
8. Teo, A. P. (2001). *Validation of an evacuation model currently under development*. University of Canterbury, School of Engineering.

9. Campanella, M., Hoogendoorn, S., & Daamen, W. (2014). Quantitative and qualitative validation procedure for general use of pedestrian *models*. In U. Weidmann, U. Kirsch, & M. Schreckenberg (Eds.), *Pedestrian and evacuation dynamics 2012* (pp. 891–905). Cham, Switzerland: Springer.
10. International Organization for Standardization. (2008). *Fire safety engineering—assessment, verification and validation of calculation methods*. ISO 16730. 2008.
11. IMO M: 1/Circ. 1238. (2007) *Guidelines for evacuation analysis for new and existing passenger ships*.
12. Ronchi, E., Kuligowski, E. D., Reneke, P. A., Peacock, R. D., & Nilsson, D. (2013). *The process of verification and validation of building fire evacuation models* (NIST Technical Note 2013).
13. Galea, E. R., Deere, S., Brown, R., & Filippidis, L. (2013). An experimental validation of an evacuation model using data sets generated from two large passenger ships. *Journal of Ship Research, 57*, 155–170.
14. Klüpfel, H. (2008). Ship evacuation—guidelines, simulation, validation, and acceptance criteria. In W. W. F. Klingsch, C. Rogsch, A. Schadschneider, & M. Schreckenberg (Eds.), *Pedestrian and evacuation dynamics 2008* (pp. 257–262). Berlin, Germany: Springer.
15. Corbetta, A., Bruno, L., Muntean, A., & Toschi, F. (2014). *High statistics measurements of pedestrian dynamics*. arXiv preprint arXiv:14071254 2014.
16. Zhang, Q., Han, B. M., & Lu, F. (2009). *Simulation model of passenger behavior in transport hubs*. 2009 International Conference on Industrial Mechatronics and Automation (pp. 220–224).
17. Schadschneider, A., & Seyfried, A. (2009). *Empirical results for pedestrian dynamics and their implications for cellular automata models*. In H. J. P. Timmermans (Ed.), *Pedestrian behavior: Data collection and applications* (pp. 27–43). Bingley: Emerald Group.
18. Shiwakoti, N., Sarvi, M., Rose, G., & Burd, M. (2011). *Animal dynamics based approach for modeling pedestrian crowd egress under panic conditions*. Papers selected for the 19th International Symposium on Transportation and Traffic Theory 2011 (Vol. 17, pp. 438–461).
19. van Hees, P. (2013). Validation and verification of fire models for fire safety engineering. *Procedia Engineering, 62*, 154–168.
20. Xu, X., Song, W. G., & Zheng, H. Y. (2008). Discretization effect in a multi-grid egress model. *Physica A: Statistical Mechanics and its Applications, 387*, 5567–5574.
21. Guo, R.-Y. (2014). New insights into discretization effects in cellular automata models for pedestrian evacuation. *Physica A: Statistical Mechanics and its Applications, 400*, 1–11.
22. Still, G. K. (2000). *Crowd dynamics*. Warwick, England: University of Warwick.
23. Helbing, D., & Molnár, P. (1995). Social force model for pedestrian dynamics. *Physical Review E, 51*, 4282–4286.
24. Song, W. G., Yu, Y. F., Wang, B. H., & Fan, W. C. (2006). Evacuation behaviors at exit in CA model with force essentials: A comparison with social force model. *Physica A: Statistical Mechanics and its Applications, 371*, 658–666.
25. Hostikka, S., Korhonen, T., Paloposki, T., Rinne, T., Matikainen, K., & Heliövaara, S. (2007). *Development and validation of FDS+ Evac for evacuation simulations*. Espoo, Finland: VTT Research.
26. Fang, Z. M., Song, W. G., Wu, H., & Zhang, J. (2012). A multi-grid model for evacuation coupling with the effects of fire products. *Fire Technology, 48*, 91–104.
27. Capote, J. A., Alvear, D., Abreu, O., Cuesta, A., & Alonso, V. (2013). A real-time stochastic evacuation model for road tunnels. *Safety Science, 52*, 73–80.
28. International Maritime Organization. (2007). *Guidelines for evacuation analyses for new and existing passenger ships*. MSC/Circ. 1238 2007.
29. Kuligowski, E. D., Peacock, R. D., & Hoskins, B. (2005). *A review of building evacuation models*. US Department of Commerce, National Institute of Standards and Technology.

30. Ma, J., Song, W. G., Zhang, J., Lo, S. M., & Liao, G. X. (2010). k-Nearest-Neighbor interaction induced self-organized pedestrian counter flow. *Physica A: Statistical Mechanics and its Applications, 389*, 2101–2117.
31. Ma, J., Lo, S. M., & Song, W. G. (2012). Cellular automaton modeling approach for optimum ultra high-rise building evacuation design. *Fire Safety Journal, 54*, 57–66.
32. Huo, F., Song, W., Lv, W., & Liew, K. (2014). Analyzing pedestrian merging flow on a floor–stair interface using an extended lattice gas model. *Simulation, 90*, 501–510.
33. Xu, X., & Song, W. G. (2009). Staircase evacuation modeling and its comparison with an egress drill. *Building and Environment, 44*, 1039–1046.
34. Predtechenskii, V., & Milinskiĭ, A. I. (1978). *Planning for foot traffic flow in buildings*. Washington, DC: National Bureau of Standards, US Department of Commerce, and the National Science Foundation.
35. Weidmann, U. (1993). *Transporttechnik der Fussgänger: Transporttechnische Eigenschaften des Fussgängerverkehrs (Literaturauswertung)*. ETH, IVT.
36. Older, S. (1968). Movement of pedestrians on footways in shopping streets. *Traffic engineering & control, 10*, 160–163.
37. Liu, S. B., Lo, S. M., Ma, J., & Wang, W. L. (2014). An agent-based microscopic pedestrian flow simulation model for pedestrian traffic problems. *IEEE Transactions on Intelligent Transportation Systems, 15*, 992–1001.
38. Zhou, J.-b., Chen, H., Yang, J., & Yan, J. (2014). Pedestrian evacuation time model for urban metro hubs based on multiple video sequences data. *Mathematical Problems in Engineering, 2014*, 1–11.
39. Zhang, J., & Seyfried, A. (2014). Comparison of intersecting pedestrian flows based on experiments. *Physica A: Statistical Mechanics and its Applications, 405*, 316–325.
40. Zhang, J., Mehner, W., Holl, S., Boltes, M., Andresen, E., Schadschneider, A., et al. (2013) *Universal flow-density relation of single-file bicycle, pedestrian and car motion*. arXiv preprint arXiv:13122735 2013.
41. Seyfried, A., Rupprecht, T., Passon, O., Steffen, B., Klingsch, W., & Boltes, M. (2007). New insights into pedestrian flow through bottlenecks. *Transportation Science, 43*, 395–406.
42. Sun, L., Yang, Z., Rong, J., & Liu, X. (2014). Study on the weaving behavior of high density bidirectional pedestrian flow. *Mathematical Problems in Engineering, 2014*, 1–9.
43. Chattaraj, U., Chakroborty, P., & Subhashini, A. (2013). Empirical studies on impacts of obstacle inside corridor on pedestrian flow. *Procedia—Social and Behavioral Sciences, 104*, 668–677.
44. Nagai, R., Fukamachi, M., & Nagatani, T. (2005). Experiment and simulation for counterflow of people going on all fours. *Physica A: Statistical Mechanics and its Applications, 358*, 516–528.
45. Seyfried, A., Steffen, B., Klingsch, W., & Boltes, M. (2005). The fundamental diagram of pedestrian movement revisited. *Journal of Statistical Mechanics-Theory and Experiment, 2005*, P10002.
46. Helbing, D., Buzna, L., Johansson, A., & Werner, T. (2005). Self-organized pedestrian crowd dynamics: Experiments, simulations, and design solutions. *Transportation Science, 39*, 1–24.
47. Mōri, M., & Tsukaguchi, H. (1987). A new method for evaluation of level of service in pedestrian facilities. *Transportation Research Part A: General, 21*, 223–234.
48. Helbing, D., Johansson, A., & Al-Abideen, H. Z. (2007). Dynamics of crowd disasters: An empirical study. *Physical Review. E, Statistical, Nonlinear, and Soft Matter Physics, 75*, 046109.
49. Hankin, B., & Wright, R. (1958). Passenger flow in subways. *Operational Research Quarterly, 9*, 81–88.
50. Bandini, S., Mondini, M., & Vizzari, G. (2014). Modelling negative interactions among pedestrians in high density situations. *Transportation Research Part C: Emerging Technologies, 40*, 251–270.

51. Fang, Z. M., Song, W. G., Liu, X., Lv, W., Ma, J., & Xiao, X. (2012). A continuous distance model (CDM) for the single-file pedestrian movement considering step frequency and length. *Physica A: Statistical Mechanics and its Applications, 391*, 307–316.
52. Schadschneider, A., Klingsch, W., Klüpfel, H., Kretz, T., Rogsch, C., & Seyfried, A. (2008). *Evacuation dynamics: Empirical results, modeling and applications*. In R. A. Meyers (Ed.), *Encyclopedia of Complexity and Systems Science* (pp. 3142–3176). New York, NY: Springer.
53. Shiwakoti, N., Sarvi, M., & Rose, G. (2008). *Modelling pedestrian behaviour under emergency conditions–state-of-the-art and future directions*. In 31st Australasian Transport Research Forum (ATRF) (pp. 457–473).
54. Helbing, D., Isobe, M., Nagatani, T., & Takimoto, K. (2003). Lattice gas simulation of experimentally studied evacuation dynamics. *Physical Review E, 67*, 067101.
55. Zhang, J., Song, W. G., & Xu, X. (2008). Experiment and multi-grid modeling of evacuation from a classroom. *Physica A: Statistical Mechanics and its Applications, 387*, 5901–5909.
56. Fang, Z. M., Song, W. G., Zhang, J., & Wu, H. (2010). Experiment and modeling of exit-selecting behaviors during a building evacuation. *Physica A: Statistical Mechanics and its Applications, 389*, 815–824.
57. Saloma, C., Perez, G. J., Tapang, G., Lim, M., & Palmes-Saloma, C. (2003). Self-organized queuing and scale-free behavior in real escape panic. *Proceedings of the National Academy of Sciences of the United States of America, 100*, 11947–11952.
58. Dussutour, A., Fourcassié V., Helbing D., & Deneubourg J-L. (2004). Optimal traffic organization in ants under crowded conditions. *Nature, 428*, 70–73.
59. Altshuler, E., Ramos, O., Nunez, Y., Fernandez, J., Batista-Leyva, A. J., & Noda, C. (2005). Symmetry breaking in escaping ants. *The American Naturalist, 166*, 643–649.
60. Ballerini, M., Cabibbo, N., Candelier, R., Cavagna, A., Cisbani, E., Giardina, I., et al. (2008). Interaction ruling animal collective behavior depends on topological rather than metric distance: evidence from a field study. *Proc Natl Acad Sci U S A, 105*, 1232–1237.
61. Shiwakoti, N., Sarvi, M., Rose, G., & Burd, M. (2011). Animal dynamics based approach for modeling pedestrian crowd egress under panic conditions. *Transportation Research Part B: Methodological, 45*, 1433–1449.
62. Shiwakoti, N., Sarvi, M., & Burd, M. (2013). Similar crowd behavior in organisms of vastly different body size. *Journal of Insect Behavior, 27*, 239–250.
63. Dias, C., Sarvi, M., Shiwakoti, N., Ejtemai, O., & Burd, M. (2013). Investigating collective escape behaviours in complex situations. *Safety Science, 60*, 87–94.

Chapter 5
Future Challenges in Evacuation Modelling

Arturo Cuesta, Orlando Abreu, and Daniel Alvear

5.1 Challenge 1: New Scenarios

5.1.1 Current Evacuation Models

To date (2014), there exist over 40 evacuation computer models. Table 5.1 shows a list of models, and their typical use/s (evacuation scenarios).

The majority of models are 'general models' that can simulate any type of building, while the "specialized models" can simulate specific scenarios (aircrafts, ships, transport stations, etc.).

Ronchi and Kinsey analysed user experience and requirements for evacuation models, via an online survey [1]. The study involved 198 participants, with different academic backgrounds, positions, and experience in the use of evacuation models. Figure 5.1 shows the proportion of participants that mainly use a given evacuation model. This study showed that 63.7 % of participants normally use 'general models' (SIMULEX [2, 3], FDS+Evac [4], VSSIM [5], STEPS [6], Pathfinder [7], building-EXODUS [8], Legion [9], and PedGo [10]). Note that EXODUS has different specialized versions (airEXODUS [11], and maritimeEXODUS [12]). These top nine 'general models' were developed and used for predicting evacuation outcomes relating to public building structures (office buildings, high-rise buildings, mass transit stations, stadiums, etc.). Other models have been developed, specifically for evacuation analysis of aircrafts and passenger ships. However, there are other scenarios where these tools have not yet been widely applied and/or validated.

A. Cuesta (✉) • O. Abreu • D. Alvear
University of Cantabria, Los Castros s/n., Santander 39005, Spain
e-mail: cuestaar@unican.es; abreuo@unican.es; alveard@unican.es

A. Cuesta et al. (eds.), *Evacuation Modeling Trends*,
DOI 10.1007/978-3-319-20708-7_5

Table 5.1 Overview of current evacuation computer models and evacuation scenarios for which they are typically used

Model name	Typical use(s)
AENEAS	Passenger ships
ALLSAFE	Any type of buildings; passenger vehicles
ASERI	Any type of buildings; passenger vehicles
BGRAFF	Any type of buildings
CRISP	Any type of buildings
DBES	Any type of buildings
E-ECAPE	Multi-storey buildings
EGRESS	Any type of buildings; transport stations; ships
EPT	Any type of buildings
EVA	Aircrafts
EVACNET4	Any type of buildings
EVACS	Any type of buildings
EVACSIM	High-rise Buildings
EvacuationNZ	Any type of buildings
Evi	Any type of buildings; maritime infrastructures
EXIT89	High-rise buildings
EXITT	Any type of buildings
airEXODUS	Aircrafts; passenger vehicles
Building EXODUS	Any type of buildings
MaritimeEXODUS	Passenger ships
FDS+Evac	Any type of buildings
GRIDFLOW	Any type of buildings
LEGION	Any type of buildings
MAGNETIC SIMULATION	Any type of buildings
MASSEgress	Any type of buildings
MassMotion	Any type of buildings
MYRIAD II	Any type of buildings
NOMAD	Any type of buildings
Pathfinder	Any type of buildings
PAXPORT/PEDROUTE	Transport stations
PEDFLOW	Any type of buildings
PedGo	Any type of buildings
SIMULEX	Buildings; transport stations
SimWalk	Buildings; transport stations
Space Sensor	Transport stations
STEPS	Any type of buildings; transport stations
TIMETEX	Multi-storey buildings
VEGAS	Any type of buildings
VISSIM/VISWALK	Any type of buildings
WAYOUT	Multi-storey buildings

Source: Evacmod web (http://www.evacmod.net/)

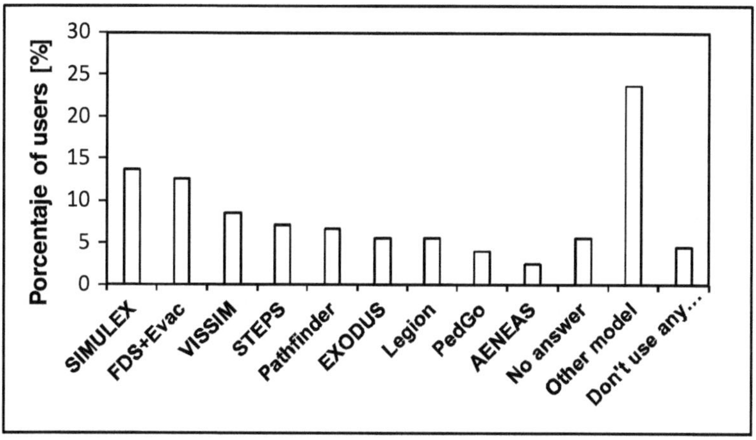

Fig. 5.1 Proportion of participants that mainly use a given model [1]

5.1.2 New Evacuation Scenarios

An evacuation scenario is any place, area, or location (facility, structure, vehicle, etc.), occupied by people, that may need to be evacuated in case of emergency. There can be macro-scale (regions, cities, etc.) and micro-scale (facilities, buildings, vehicles, etc.) evacuation scenarios. Here we focus on micro-scale evacuation scenarios. The term "evacuation scenario" includes the physical and social environment, as well as the circumstances around the occupants. It is evident that each evacuation scenario is unique. This term defines both the actual and the simulated conditions.

The PD7974-6:2004 document provides a list of evacuation scenarios for fire safety engineering analyses [13]. The list includes:

- A—Office or industrial.
- B1—Shop, restaurant, circulation space.
- B2—Cinema, theatre.
- Ci—Dwelling.
- Cii—Service flats, halls of residence, etc.
- Ciii—Hotel, hostel.
- D—Residential (institutional/medical care).
- E—Railway station/Airports.

It can be said that the 'general models' are suitable for scenarios in categories A, B1, B2, Ciii, and E (see Fig. 5.1). These evacuation scenarios involve, in most cases, the simulation of self-rescue processes (e.g., occupants respond and move to a place of safety by themselves). However, other evacuation scenarios have not yet been studied in depth. It is difficult to apply current evacuation models to some of them, and some may require a specialised model. For example, scenarios in categories Ci,

Cii, and D may involve assisted evacuation, due to the presence of vulnerable people who cannot evacuate by themselves (hospitals, day-care centres, kindergartens, schools, etc.). Few evacuation models are able to simulate assisted evacuations. The EXITT model includes two types of occupant, those fully capable and those who are in need of assistance to evacuate the building. Decision rules apply only to the first group, and the latter group follows the decisions and movements made by their rescuers [14]. The BUMMPEE model can simulate the evacuation of people with disabilities, and their interaction with the environment [15]. However, it is not clear whether the model can simulate assisted evacuation.

There are two models that specialise in the evacuation of hospitals: the G-HES [16] and the Assisted Evacuation Simulation System [17]. Both models incorporate characteristics of assisted evacuations. However, to the author's knowledge, they have not been validated and are not publicly available. Given this, one alternative is the use of a current evacuation model. Some 'general models' allow the simulation of additional behaviours, such as travel itineraries assigned to occupants. This could be used to simulate prescript assisted evacuation procedures. In a study for the Fire Protection Research Foundation, the STEPS model was used to simulate the relocation of patients on a sleeping room floor in a health care facility [18]. The report highlighted the lack of empirical data and the limitations of using a 'general model' for these kinds of scenarios, and suggested future development for addressing the problem of simulating assisted evacuation.

Other scenarios for which it is difficult to apply current evacuation models include those involving children (e.g., school buildings). This is because, although the majority of 'general models' are able to represent heterogeneous populations, this is not supported by empirical data, leading to limitations in predicting an evacuation involving children.

Some research works have been conducted in order to provide new data. Most of these studies present information relating to the walking speeds (horizontal and on stairs) of children, according to the age [19–22]. Attempts to simulate the evacuation of children using 'general models' are scarce. Rosaria Ono et al. carried out one of these studies [23]. This work mainly focused on improving the design of school buildings by using the buildingEXODUS model, while using the default data from the Simulex model to represent the walking speed of children (0.90 ± 0.30 m/s). Campanella et al. compared the results of three evacuation drills in day-care centres (children 0–2 and 3–6 years old) with the pedestrian simulation Nomad [24]. This model underestimated the total evacuation times in two out of three cases, due to occurrences which could not be simulated, such as doors that could not be opened by children. In another contribution, empirical observations from an evacuation drill in a school building were compared with the simulation results from STEPS, GridFlow, and Pathfinder [25]. Results highlighted the risk of using default values, leading to errors in evacuation time predictions of over 20 %. One of the most recent contributions explored the capabilities of STEPS and Pathfinder for simulating the evacuation of children through stairs [26]. This study used empirical data for pre-schoolers and secondary school children. The results suggested the need to take into account other input parameters, in addition to walking speeds, in order to obtain accurate predictions. These include the "maximum room density" for Pathfinder, and the flow rates ("stair exit capacities") for STEPS.

Category E includes railway stations, airports, terminals, etc. However, there can also be other new scenarios in transportation, such as passenger trains and road tunnels.

There are two main methods for the analysis of evacuations from rail vehicles: First, evacuation drills and/or egress experiments and, second, the use of modelling and simulation. Data from evacuation drills and experiments are scarce, and mainly related to exit flow rates in a limited number of evacuation scenarios (evacuation of upright vehicles to platform and rail level) [27–32]. There are two exceptions. The first presents data from evacuation experiments in an overturned rail car in smoky conditions. However, due to the limited number of tests, results are more indicative than definitive [33]. The second describes the strategies adopted by passengers when evacuating to the rail level [34]. Modelling and simulation has been in use for last few years. The first studies were related to fire safety engineering. In one study conducted by the NIST (National Institute of Standards and Technology), calculations were performed in order to estimate the minimum necessary egress time of an upright rail car with a maximum capacity of 72 passengers [35]. As the report states, different assumptions were considered, showing the lack of empirical data and the absence of a specific tool that permits the simulation of passengers performing additional activities, such as collecting belongings, investigating the fire, etc., instead of simply exiting the car. Another contribution was carried out by the Department of Fire Safety Engineering (Lund University), who published a report about safety conditions in the case of a fire in an intercity train [36]. However, many assumptions were made regarding the characteristics and passenger behaviours, and the model showed limitations. Although these studies analysed simple scenarios, many assumptions were necessary, due to the lack of data and the limitations of the evacuation models (of which none were designed for rail vehicles) when applied to rail vehicles. Capote et al. presented a new simulation approach in order to capture the impact of passenger behaviours and activities inside the passenger compartments when an evacuation becomes necessary. This study used empirical data for passenger performances in evacuation trials. The resulting model is a specific evacuation model for high-speed passenger cars [37, 38]. The FRA (Federal Railroad Administration) sponsored a research program, to investigate the potential applicability of performance-based criteria to the evacuation times of rail vehicles in the USA. Within this research program, the Volpe Center investigates the feasibility of using modelling and simulation. The study concludes that current models (mainly designed for buildings) are not appropriate for simulating evacuation from rail vehicles [39]. The study incorporates the development of a prototype of a passenger rail car computer model. The model is a modified version of buildingEXODUS, with the capability to simulate egress from rail cars to high and low platforms [40].

Road tunnels are particular scenarios where the application of current evacuation models can also be questionable. Shields and Boyce stated that the understanding of human behaviour during a fire in buildings might not be readily transferable/applicable to human behaviour in response to fire in tunnels [41]. Norén and Winér developed a conceptual model for human response in road tunnels in cases of emergency based on observations from evacuation tests in the Benelux Tunnel [30]. The analyses showed that the evacuation process involves three stages: 1) time inside the car, 2) hesitation time, and 3) movement time. However, the observations showed that the behaviour of the motorist did not fit precisely to the three-stage model. There were two different groups of motorist, those leaving their cars before and after the announcement, respectively. Due to this, the hesitation duration differed highly between the motorists, and some of them did not hesitate after leaving their cars. Attempts to simulate the evacuation of road tunnels using 'general models' are scarce. Ronchi et al. compared three building evacuation models (FDS+Evac, STEPS, and Pathfinder) in a case study of a road tunnel [42]. The results showed substantial differences between the models. Neither STEPS nor Pathfinder could reproduce the environmental conditions caused by the smoke, while this affected the choice of a particular exit in FDS+Evac. EvacTunnel 3.0 is an example of a specialized evacuation model for road tunnels [43]. This model for risk analysis applications includes a stochastic approach, individual perspective, and fire effects on evacuation. The model is available at (http://www.gidai.unican.es/).

5.1.3 New Scenarios and New Models

Evacuation modelling, as other modelling activities, has a basic principle: the clear definition of the overall objectives of the simulation, their purpose, and the specific scenario of application.

There are three possible alternatives for the analysis of a new evacuation scenario: 1) select and use a 'legacy' model, 2) improve a current evacuation model, or 3) develop a new evacuation model.

The first option consists of the selection of an evacuation model that has been developed either in the past, or for a different evacuation scenario. This is not an easy task, because several evacuation models exist, which vary in their background, capabilities, characteristics, sophistication, etc. Other than the possibilities of accessing a specific model (availability, cost, etc.), there are other criteria in the selection process. A potential user may obtain information from the evacuation model reviews that are available in the literature. The SFPE handbook of Fire Protection Engineering proposes three approaches to evacuation modelling: 1) optimization, 2) simulation, and 3) risk assessment [44]. Kuligowski and Peacock provide information for building evacuation models and their features (availability, modelling method, purpose, structure, perspective, etc.) [45]. According to Gwynne et al., evacuation models fall into two categories: those that only consider human movement, and those that attempt to link movement with behaviour [46]. Santos and Aguirre propose four categories: 1) flow based, 2) cellular automata, 3) agent-based, and 4) activity-based models [46], while Tabares [3] considers three groups: 1) a macroscopic approach, 2) a microscopic approach, and 3) an effect-based simulation approach [47]. This information can be useful for knowing the capabilities, and the levels of sophistication, of different evacuation models. However, model reviews do not provide any criteria for the selection of an evacuation model. There are few references in existing literature regarding this problem. Nelson & Mowrer provide basic questions that a potential model user should consider when selecting a model [44]. The questions focus on the model's capabilities and sophistication, regarding the representation of a given scenario. Kuligowski and Gwynne provide the key factors to consider for the selection of an evacuation model [48]. These factors focus on the project requirements, and the model characteristics (background and sophistication).

We propose here a qualitative method of providing criteria for the selection of a model, for an application to a new evacuation scenario. Note that the best criterion for the selection and use of an evacuation model is validation (see Chap. 4). Despite this, the use of a model beyond its validation evidence requires an additional effort, in order to understand the limitations of a model in representing the evacuation process of a new scenario. The first step is to assess the fidelity of the model, defined as the degree to which the model represents the state and behaviour of the real world scenario in a measurable or perceivable manner. For this, it is necessary to select the aspects of the scenario that are essential for the model to represent, while excluding those aspects that are not relevant for the purpose of the simulation.

These essential aspects relate to: 1) a reference scenario (RS), 2) a reference population (RP), 3) a changing scenario (CS), and 4) a changing population (CP). The RS and RP constitute the initial conditions of the physical and human resources, respectively, and they are defined by some initial parameters.

The CS and CP represent the change in the initial conditions of the physical and human resources, respectively, and they may vary over time. These changes depend on the development of the emergency and evacuation processes. Here we use the Propositional Logic (PL) language:

$$F = F_{RS} \wedge F_{RP} \wedge F_{CS} \wedge F_{CP} \tag{5.1}$$

The F_{RS} term represents the fidelity of the model for the mathematical reproduction of the physical conditions of the scenario, such as the space discretization (fine or coarse network, or continuous space), and other elements such as escalators, turnstiles, obstacles, vehicles, etc. This depends on the model perspective of occupants (individual and/or global), and the movement method. The F_{RP} term represents the fidelity of the model to represent the number, characteristics, and distribution of the population (occupants or groups of occupants). This depends on the model perspective and the behavioural method. The F_{CS} term represents the fidelity of the model to represent changes in the reference scenario, due to the emergency. The scenario can change in a certain moment (e.g., an explosion, earthquake, collapse, etc.), or gradually (a fire, flood, etc.).

The most complex is the capability of the model to represent the changes in the scenario during the course of the simulation, and their potential impact on the population and, therefore, the evacuation process. The F_{CP} term represents the fidelity of the model to represent the occupant responses and actions during the course of the simulation. If the selected model does not meet an appropriate fidelity (F) in one these essential aspects (F_{RS}, F_{RP}, F_{CS}, F_{CP}), then we can reject it for the intended use, because it is likely not to satisfactorily represent the real phenomenon in question. Nevertheless, how do we know that?

An example of the fidelity assessment for the changing population (CP) can be given as follows:

$$\bar{F}_{CP} = I \wedge B \tag{5.2}$$

$\bar{F}_{CP}$ is the statement 'the model does not have the appropriate fidelity to represent the changing population', I is the statement 'the scenario needs an individual perspective of occupants rather than a global perspective of occupants', and B is the statement 'the observed behaviour of the occupants does not fit to the scheme pre-movement>movement'. The statement B is the key condition for knowing if the model in question has the required fidelity for the CP. When $\bar{F}_{CP} = True$, the model should not be used for the new scenario.

The I statement can be formulated by the following expression:

$$I = \frac{\left| t_{evac1} - t_{evac2} \right|}{t_{evac2}} \geq \delta_{acept} \tag{5.3}$$

where

t_{evac1}—Total evacuation time obtained from a global perspective of occupants;

t_{evac2}—Total evacuation time obtained from an individual perspective of occupants;

δ_{acept}—Acceptable level of the relative error.

Equation (5.3) represents the relative error of the evacuation times obtained from both the individual and global perspective of occupants. This equation assumes the hypothesis that the evacuation time calculated from an individual perspective is closer to the real one.

The acceptable level of the relative error depends on the problem at hand, and the needed accuracy. Lower the value of this acceptable level, greater the accuracy demanded. The proposed values could be 0.05, 0.10, or 0.15. The evacuation times from both the individual and global perspectives of occupants can be calculated manually in a preliminary stage in accordance with the equations (5.4) and (5.5), respectively [3]:

$$t_{evac1} = t_{pre(1st\ perc)} + t_{trav} + t_{flow} \tag{5.4}$$

where

$t_{pre(1st\ perc)}$—Time from the alarm to the movement of the first few occupants (percentile 1);

t_{trav}—Walking time to the exits (a conservative estimate could use the maximum direct travel distance);

t_{flow}—Time for total occupant population to flow through the available exits.

$$t_{evac2} = t_{pre(1st\ perc)} + t_{pre(99th\ perc)} + t_{trav} \tag{5.5}$$

where

$t_{pre(99th\ perc)}$—Time from the alarm to the movement of the last few occupants (percentile 99);

When $I = True$, an individual perspective should be used for the scenario. Many of the existing evacuation models use the individual perspective of occupants. However, this is not sufficient, as a necessary second step consists of checking whether the selected evacuation model has the capability to represent the observed evacuation behaviour of occupants in the new scenario. This is represented by the propositional variable B of (5.2). A computer model that attempts to simulate evacuation behaviour must be able to predict two things: first, the actions that (simulated) occupants perform, and second, how long it takes to perform each action [9]. To date, evacuation models can represent the evacuation behaviour through two types of variables: time variables (t), and movement variables (m).

Time variables are those variables that are not related to purposive evacuation movement expressed in delays due to different reasons (response, seeking information, collecting belongings, waiting, getting dressed, etc.). Movement variables are those variables related to purposive evacuation movement (travel times). The modelled evacuation behaviour commonly follows the sequence of Fig. 5.2, by assuming a time in which occupants will wait in their initial position (time variables), before beginning evacuation movements (movement variables). Furthermore, the period before the evacuation movement is usually represented by a unique variable (called pre-movement, pre-evacuation time, response time, etc.). This is common in most building scenarios, transport terminals, etc. However, some new evacuation scenarios may not fit the schema of Fig. 5.2.

Let us consider the evacuation of a school building. The alarm sounds. One of the classrooms on the first floor takes a certain time to prepare for the evacuation (children form a queue and wait for the teacher's instructions to start the evacuation). This can constitute the first time variable (t_1). Subsequently, the teacher orders the evacuation, and the children move towards the stairs. This can constitute the first movement variable (m_1). Then, due to the evacuation procedure, the teacher and the

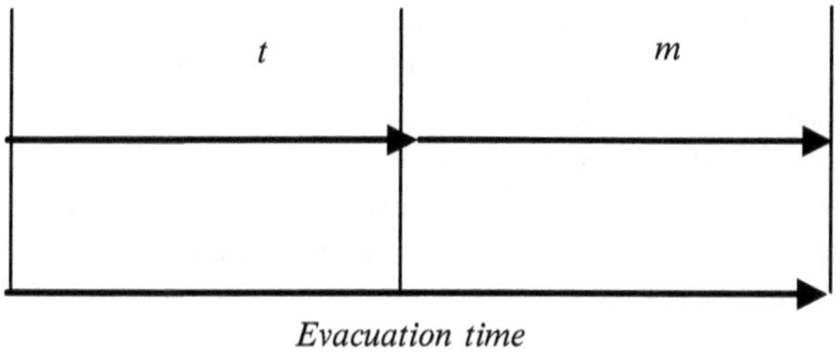

Fig. 5.2 Schema of the evacuation behaviour commonly used by current evacuation models

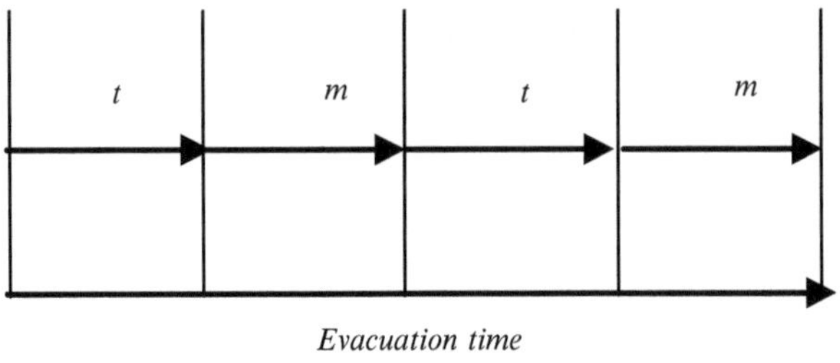

Fig. 5.3 Schema of the evacuation behaviour observed in a school building

children wait until the stairs, used by another classroom, get clear. This can constitute the second time variable (t_2). Next, the teacher and the children walk down the stairs. This can be represented by the second movement variable (m_2). The schema of the described evacuation behaviour is represented in Fig. 5.3.

If we use the schema of Fig. 5.2 to represent the evacuation behaviour described above, it would be necessary to represent t_1 and t_2 in a unique time variable.

The following two questions should be answered here. Can we represent all of the time variables with only one variable? Can we omit interruptions in the evacuation movement that are caused by certain interactions and behaviours? Regarding the first question, by taking into account the fact that t_1 and t_2 are two random variables, each with a mean and a standard deviation, they should fit to the same probability distribution law. Furthermore, the relations from (5.6) and (5.7) must be satisfied. Otherwise, the model will provide an inaccurate representation of reality, leading to errors and uncertainty in the results.

$$\overline{t} = \overline{t_1} + \overline{t_2} \tag{5.6}$$

$$\sigma_t = \sqrt{\sigma_{t_1}^2 + \sigma_{t_2}^2} \tag{5.7}$$

Regarding the second question, t_2 does not arise from physical constraints (congestion, bottleneck), and it is not simply an additional time without movement. Its position in the timeline of evacuation behaviour is important, because it affects the preceding and following variables. If we take the aforementioned example of the evacuation in the school building. The omission of t_2 means direct access for the teacher and the children to the stairs. This leads to an underestimation of the evacuation time for the first floor, and a simulation of movement conditions (flow, density) on the stairs that are far from the actual process.

There are several situations and scenarios where evacuation behaviour schemas may differ from the one in Fig. 5.2. For example, scenarios where B is True (see (5.2)). In that case, B is the propositional variable that takes into account the behaviour characteristics of occupants.

$$B = B_1 \wedge B_2 \tag{5.8}$$

where B_1 characterizes the evacuation behaviour of each occupant, and B_2 is the weight of B in the total occupant population of the evacuation scenario. Then, B_1 can be formulated as follows:

The behaviour of occupants is represented in tuples Z_i:

$$Z_i = \left(z_{i1},,, z_{i2},,, \ldots,,, z_{ini} \right) \tag{5.9}$$

This constitutes an n_i-tupla with parameters of i-th occupant, taking into account:

$$z_{ij} = t_{ij} \wedge m_{ij} \tag{5.10}$$

where

t_{ij}—The j-th time variable (delays, actions) of the i-th occupant:

m_{ij}—The j-th movement variable (travel speed, flow) of the i-th occupant;

$i=[1,n_{p0}]$

$n_p o$—The number of occupants in the scenario;

$j=[2,n_i]$

n_i—The number of variables for the i-th occupant. There must be at least two variables: one time variable and one movement variable.

B_1 describes the following relation of variables:

$$B_1 = \left(z_{ij} = m_{ij} \right) \wedge \left(z_{ij+1} = t_{ij+1} \right) \wedge \left(j+1 < n_i \right) \tag{5.11}$$

Given the number of occupants with the relation described above:

$$n_{B1} = \sum_{i=1}^{n_p 0} B_1(i) \tag{5.12}$$

where

$$B_1(i) = \begin{cases} 1 - if \;\; B_1 = True \;\; for \;\; i-th \;\; occupant \\ 0 - if \;\; B_1 = False \;\; for \;\; i-th \;\; occupant \end{cases}$$

B_2 can be formulated as follows:

$$B_2 = n_{B1} \geq 0.2 n_{p0} \tag{5.13}$$

Equation (5.13) represents the minimum proportion of occupants with evacuation behaviours that are different from the conventional schema (see Fig. 5.2). When F_{CP} is true, "the model does not reproduce the observed reality" of the new scenario.

In this case, there are two further options: 1) improve the current evacuation model and validate it for the new scenario, and 2) develop a new evacuation model that specialises in the new scenario, and validate it.

In majority of cases, the improvement of current evacuation models is restricted to the developers because most of the software is closed-source (the end-users cannot access the source code). Kuligowski and Gwynne stated the importance of flexibility and long-term capabilities in evacuation models for future improvements in addressing new emergency scenarios, procedural evacuations, interaction of the model with technology, real-time manipulation, etc. [48].

Some of the current models are flexible enough and are constantly improved in order to address new end-user demands [48].

The last option is the development of a specialised model, tailored to the project requirements and the objectives of the simulation. These models should be conceived as research tools with the purpose of addressing a specific and well-defined problem. Note that the development of conceptual, mathematical, and computational models is not necessarily a tedious task when high visualization quality is not required in the model.

A specialised model will have no further use then for its domain. It will be developed with one purpose. Therefore, it is argued here that these types of models can provide more consistent and reasonable results. The authors recommend the development and use of stochastic evacuation models in the future, rather than deterministic approaches. In the next section, we propose two methods in order to assess whether a particular scenario should be studied using a stochastic or a deterministic evacuation model.

5.2 Challenge 2: Stochastic vs. Deterministic

Evacuation is essentially a stochastic process. Despite this, the majority of current evacuation models tackle the phenomenon in a deterministic way. Some models use distribution laws to treat the randomness of human actions and decision inputs; however, they essentially use deterministic approaches (e.g., use relations between speed and density, produce only one sample of possible results, etc.). Averill expressed the importance of a stochastic approach [8, 9] in evacuation analysis in [49]. Lord et al. discussed the role of distributions in modelling inputs and outputs, for addressing the uncertainty [50]. However, it is clear that this approach is more complex than deterministic modelling, in terms of data collection and processing, computational sophistication and run times, output processing, etc. We propose the following scenario-dependent method that could aid researchers and designers in choosing the correct modelling approach for the future.

5.2.1 Determinism or Randomness: The Exact Method

All processes in nature are essentially random [51], because they are based on a set of unpredictable factors that cause the results to vary according to chance. However, in some cases this influence can be minor, or even imperceptible. We can obtain a mathematical representation of these circumstances. Let us consider the phenomenon's result as a set of variables:

$$Y = \{\bar{y}_i + \varepsilon_i\}\big| i = 1, n \tag{5.14}$$

where

$\bar{y}_i$—The mean value of variable Y_i;

ε_i—The random fluctuation of variable Y_i;

n—The total number of variables that characterize the phenomenon or process.

For practical purposes, if we assume a deterministic approach, the phenomenon would be defined by average characteristic variables. Therefore, a relative error would be assumed:

$$[\delta] = \{\delta_i\} = \frac{|\varepsilon_i|}{\bar{y}_i}\bigg| i = 1, n \tag{5.15}$$

In order to assume a deterministic approach, we should consider an acceptable relative error $\tilde{\delta}$. It should be noted that the acceptable $\tilde{\delta}$ can be different for each of the parameters that characterise an evacuation process. However, we will assume the same acceptable $\tilde{\delta}$ for all of these variables. Therefore, the deterministic approach is accepted if

$$\forall\ \delta_i \to \delta_i \le \tilde{\delta}\big| i = 1, n \tag{5.16}$$

Nevertheless, we cannot ignore the fact that ε_i is a random parameter. This implies that the inequality (5.16) has a casual character, and its probability of occurrence should be considered:

$$\Pr\left[\forall\ \delta_i \to \delta_i \le \tilde{\delta}\right] = p_o \big| i = 1, n \tag{5.17}$$

where

p_o—The probability of occurrence (a high probability is assumed 0.9, 0.95, 0.99…).

By considering the deterministic character of $\bar{y}_i$ and δ, and (5.15) and (5.17), we can obtain that

$$\Pr\left[\delta_i \le \tilde{\delta}\right] = \Pr\left[\frac{|\varepsilon_i|}{\bar{y}_i} \le \tilde{\delta}\right] = p_o \Rightarrow \frac{\max p_o |\varepsilon_i|}{\bar{y}_i} \le \tilde{\delta} \tag{5.18}$$

where

$\max_{p_o} |x|$—The maximum value with probability P_0 of x.

The maximum value of ε_i may be represented by

$$\max_{p_o} |\varepsilon_i| = \begin{cases} P_{p_o}(y_i) & si \quad \varepsilon_{i_{crit}} > 0 \\ \bar{y}_i - P_{1-p_o}(y_i) & si \quad \varepsilon_{i_{crit}} < 0 \end{cases} \tag{5.19}$$

where

p_p—The percentile of the random variable x;

$\varepsilon_{i_{crit}}$—The critical value for the random fluctuation of the variable y_i. Based on the variable application, the effectiveness of the phenomenon is improved when ε_i increases, or vice versa.

Equations (5.18) and (5.19) indicate that the acceptance of a deterministic approach depends on the equation compliance

$$\begin{aligned} &\frac{P_{p_o}(y_i) - \bar{y}_i}{\bar{y}_i} \le \tilde{\delta} \quad si \quad \varepsilon_{i_{crit}} > 0 \\ &\frac{\bar{y}_i - P_{1-p_o}(y_i)}{\bar{y}_i} \le \tilde{\delta} \quad si \quad \varepsilon_{i_{crit}} < 0 \end{aligned} \tag{5.20}$$

In evacuation modelling, the main output variable that characterizes the evacuation process and modelling result is the total evacuation time T_{evac}. This variable permits us to define whether people are able to leave the area safely within a reasonable time, in a specified scenario and hazardous situation. The acceptance or rejection of a deterministic approach in this kind of process has critical consequences for the saving of lives.

From the previous equations, and by considering that for this modelling purpose, if T_{evac} increases, the situation is less favourable (condition $\varepsilon_{i_{crit}} > 0$); we assume that a deterministic approach can be acceptable if

$$\delta(T_{evac}) = \frac{P_{p_o}(T_{evac}) - \bar{T}_{evac}}{\bar{T}_{evac}} \le \tilde{\delta} \tag{5.21}$$

where

$\bar{T}_{evac}$—The mean value of evacuation time;

$P_{p_o}(T_{evac})$—P_O percentile of evacuation time.

Equation (5.21) permits us to define whether a deterministic approach can be used for a specific evacuation model in a particular scenario. In order to use (5.12), $\tilde{\delta}$ should be defined; however, this acceptable error is commonly assumed to be 0.05, 0.1, or 0.15 for engineering applications. Furthermore, the mean and P_O (i.e. 0.99) percentile values of the evacuation time should be obtained.

Several evacuation drills should be performed in order to obtain the mean and the percentile of evacuation times experimentally. Furthermore, these drills will obtain uncertain results, since it is unlikely that the real conditions of an emergency are represented in an evacuation drill. A stochastic model for the application scenario should be developed in order to obtain the mean and percentile values of the evacuation time. This requires mathematical and computational modelling of the dependency

$$\overline{T}_{evac} = f(x_j)\big| j = 1, m \quad (5.22)$$

There are different types of independent variables x_j in an evacuation model, depending on the selected modelling features (type of network, movement representation, behavioural characteristics, etc.), application scenarios, and modelling purpose. The randomness of the variables x_j will define the randomness of dependent variable T_{evac}. Equation (5.22) allows one to obtain a stochastic evacuation model. The probability distribution laws of x_j (and its statistical characteristics) allow one to obtain a sample of evacuation times using Monte Carlo methods [52, 53].

The statistical treatment of the results of the sample enables one to obtain its probability distribution law and statistical parameters (mean and percentile). Finally, (5.21) analyses the convenience of using a deterministic instead of a stochastic approach. It should be noted that a stochastic model can be used as a deterministic model by considering the dispersion (variance of standard deviation) of independent variables near zero. The hypothesis that the greater the randomness of the independent variables x_j, the greater the randomness of the dependent variable T_{evac}, can be assumed, and a stochastic approach should be considered for evacuation modelling. Therefore, if we can evaluate the level of randomness of the independent variables x_j, we will be able to define a priori the convenience of using a stochastic approach, without obtaining the complete mathematical and computational models. This is explained in the next section.

5.2.2 Determinism or Randomness: A Priori Method

The a priori method analyses the level of randomness of the independent variables for a hypothetical evacuation model. This allows the inference of the convenience of using a stochastic approach. It should be noted that this method yields estimated, rather than conclusive, results. The impact of the independent variables' randomness on dependent variable randomness (see (5.22)) is determined by the individually sensitive correlation between the independent variables, and the functional compensation between them. However, this could offer an a priori assessment of the convenience of stochastic modelling.

The level of randomness of an independent variable may be analysed by measuring the dispersion of this variable, in relation to its mean value. This can be achieved with the use of relative magnitudes:

$$\xi_{x_j} = \frac{x_j - \overline{x_j}}{\overline{x_j}} \quad (5.23)$$

where

ξ_{x_j} —The level of randomness of variable x_j;

$\overline{x_j}$ —The mean value of x_j.

It should be noted that ξ_{x_j} is a function of the random variable x_j and, therefore, similarly to (5.17), the probability of its occurrence is

$$\Pr\left[\frac{x_j - \overline{x_j}}{\overline{x_j}} < \xi_{x_j}\right] = p_o \tag{5.24}$$

where

ξ_{x_j} —The acceptable level of randomness of variable x_j.

Equation (5.14) is similar to Chebyshev's inequality [54], which is accepted for any distribution law of an independent variable x_j.

By considering the central limit theorem (CLT) [55], which states that, given certain conditions, the mean of a sufficiently large number of independent random variables will be approximately normally distributed, (5.14) can be compared with:

$$\Pr\left[\frac{x_j - \overline{x_j}}{\overline{x_j}} < kC_v\left(x_j\right)\right] = erf\left(\frac{k}{\sqrt{2}}\right) \tag{5.25}$$

where

$C_v\left(x_j\right) = \frac{\sigma_{x_j}}{\overline{x_j}}$ —The coefficient of variation for x_j;

σ_{x_j} —The standard deviation of x_j;

k—A positive number;

$erf\,(x)$—The error function.

By comparing (5.24) and (5.25),

$$p_o = erf\left(\frac{k}{\sqrt{2}}\right) \tag{5.26}$$

Furthermore, by setting P_0=0.99 in (5.26),

$$k = 2.576 \tag{5.27}$$

Equation (5.18) can be obtained by comparing (5.24) and (5.25), considering (5.27), and extrapolating the inequality in (5.24):

$$C_v\left(x_j\right) \le 0.388\xi_{x_j} \tag{5.28}$$

The definition of a level of randomness is a highly refutable task, and depends on several factors; however, we assume the value ranges shown in Table 5.2.

The levels of randomness defined in Table 5.2, and expression (5.28), establish the acceptance criteria (estimation) for using a deterministic approach in evacuation modelling (see Table 5.2). In order to accept (or reject) a deterministic approach, all

Table 5.2 Acceptable levels of randomness

Acceptance	ξ_{x_j}	Meaning
Acceptable	0.1	The variance varies no more than 10 % of its mean value with a probability of 0.99
Imprecise	[0.1, 0.2]	The variance varies between 10 and 25 % of its mean value with a probability of 0.99
Non acceptable	0.25	The variance varies no less than 25 % of its mean value with a probability of 0.99

Table 5.3 A priori acceptance criteria for a deterministic approach

Acceptance of a deterministic approach	Condition
Acceptable	$\forall x_j \mid j = 1, m \rightarrow C_v(x_j) \leq 0.0388$
Imprecise	$\forall x_j \mid j = 1, m \rightarrow 0.0388 < C_v(x_j) < 0.097$
Non acceptable	$\forall x_j \mid j = 1, m \rightarrow C_v(x_j) \geq 0.0097$

of the independent variables should satisfy (or not) the conditions established in Table 5.3.

To sum up. Two different methods have been proposed in order to analyse the convenience of using either a stochastic or a deterministic approach in evacuation modelling. The exact method can be applied by checking (5.17), and the a priori method is based on the acceptance criteria presented in Table 5.2.

5.3 Challenge 3: New Applications

Evacuation models have been widely applied for performance-based assessments of buildings, as well as other common evacuation scenarios (ships, rail stations, stadiums, etc.). They have also been applied to forensic analyses. However, there is an expansion of the opportunities for evacuation modelling, and new applications are demanded. Gwynne and Kuligowski present a classification of the different application modes (degrees of simulation): Naïve, Operational, Predictive, Engineered, Interactive, and Real-time [56]. These application modes require different levels of data and user expertise. Naïve and Operational are used for normal conditions, while the rest are used in cases of evacuation. They propose, as a future goal, the development and use of a model capable of integrating all of the application modes. This model would have the capability to operate in real-time, during both normal and emergency conditions, and provide the staff and the operator with information for dealing with different situations.

The possibility of analysing and predicting the impact of different evacuation scenarios and procedures in real-time is relevant for the decision-making process during an actual emergency. This is one of the most important future applications for evacuation modelling. Cuesta et al. introduced this concept, and proposed two evacuation models supporting decisions during actual emergencies [57]. The models have the capability to perform several simulations, by changing key random parameters and capturing all potential outcomes, and provide results within a few seconds. An additional approach to this concept is currently adopted in the GETAWAY Project [58]. This project aims to design an intelligent emergency signage system, for the real-time direction of transport terminal passengers during an evacuation. This intelligent system includes evacuation simulation; however, to date it is not clear how the evacuation model that is used performs the simulations and provides results (faster than real-time).

Indeed, making decisions is the most important task for a safety manager (operator), and is often very difficult. Past incidents have demonstrated the impact of making wrong or late decisions in cases of emergency (hurricanes, earthquakes, terrorist attacks, fire incidents, etc.). Decision support systems (DSSs) have been applied to different situations and scenarios [59, 60]. These tools may offer a more appropriate course of action and decisions, increasing the rapidness and efficiency of the user's response [61].

We define "emergency scenario" as the set of human and physical resources—buildings, structures, roads, transportation, geographical and climatic factors, etc.—that can interact with each other, and are modified or transformed during an emergency or evacuation process. As for the evacuation scenario described in Sect. 5.1, the emergency scenario comprises the following: 1) a reference scenario (RS), 2) a reference population (RP), 3) a changing scenario (CS), and 4) a changing population (CP). The RS and RP are the initial conditions defined by some initial parameters.

The CS and CP represent the change in the initial conditions that may vary over the time. These changes depend on the development of the emergency and evacuation processes.

The challenges involved are as follows: 1) assessing the convenience and usefulness of integrating an evacuation model into a DSS; 2) establishing recommendations for the performance features of the model, 3) determining the type of scenarios and emergencies where an evacuation model can be useful, and 4) defining the possible decisions for the integrated system. In order to address these issues, the following questions should be answered:

- Which performance characteristics should be used for the evacuation model?
- Which outputs from the evacuation model can be used for decision-making?
- What additional information about the emergency process is required for decision-making?
- What types of decisions can be proposed?
- Under what scenarios is it appropriate to apply the integrated systems?
- What are the methodological principles of this integration?
- How much influence does the available safe egress time (ASET) have on decision-making?

5.3.1 Characteristics of Evacuation Models for Real-Time Decision-Making

An appropriate computational run time is required for the use of an evacuation model in real-time decision-making. The total time during which the system is running or executing (run time) includes the run times of: 1) the evacuation model (T_{ModRun}), and 2) other elements of the DSS ($T_{DSSrest}$). As (5.29) shows, the total run time needs to be shorter than the acceptable time for decision-making ($T_{AccepDec}$).

$$T_{ModRun} \leq T_{AcceptDec} - T_{DSSrest} \tag{5.29}$$

The acceptable time for decision-making ($T_{AccepDec}$), depends on the scenario. This can vary from seconds (or minutes), to days. For example, in a building fire, decisions should be made within a few seconds, in order to minimize any possible consequences. In other cases, such as natural disasters (hurricanes or floods), existing forecasting tools allow for a longer acceptable time period (i.e. one or two days) for decision-making.

Current evacuation models are able to run within a few seconds, regardless of the type of grid (fine, coarse, or continuous), pedestrian movement (agent-based, cellular automata, etc.), or modelling approach (deterministic or stochastic).

It should be noted that for the model to run in seconds, the following assumptions should be considered: 1) the main input variables are already introduced in the evacuation model, 2) visualization is not used, and 3) the RP (Reference Population) may already be included and/or constantly be updated.

The evacuation time T_{evac} is usually the main output. This parameter provides useful information to the emergency manager, for establishing appropriate procedures to evacuate the affected area. Note that T_{evac} may include the total evacuation time, the evacuation times for each occupant, evacuation times for different routes, the evacuation time to clear a determined area inside the scenario, etc.

Figure 5.4 shows the well-known time components of an evacuation process based on the analogy of the fire phenomenon represented in Technical Report ISO/TR 16738:2009(E) 'Fire-safety engineering—Technical information on methods for evaluating behaviour and movement of people' [62]. As we can see in Fig. 5.4, the time of the emergency is the start of the emergency event. Then, the time at which people should be warned to evacuate, $t_{alarm,}$ must be determined. Before people start moving (evacuating), they will engage in other activities. This corresponds to a pre-evacuation time t_{pre}. Finally, people take time (movement time t_{move}) to reach a place of safety through an egress route. The evacuation time consists mainly of the pre-evacuation and movement times:

$$t_{evac} = t_{pre} + t_{mov} \tag{5.30}$$

The terms ASET and required safe egress time (RSET) are well known [10–12]. The RSET (t_{RESET}) has been widely studied, using existing evacuation models [13].

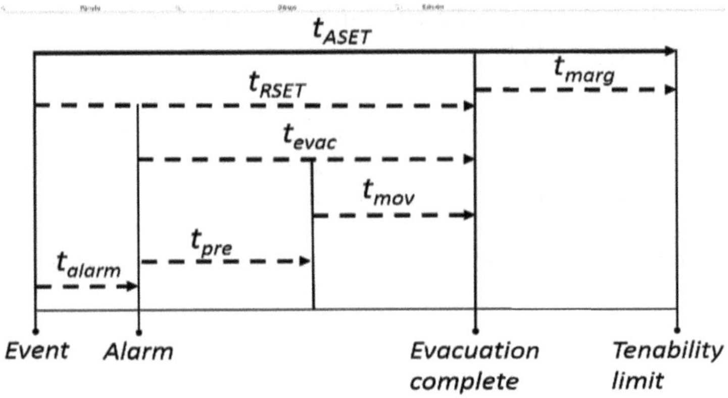

Fig. 5.4 Times in overall evacuation process

$$t_{RSET} = t_{alarm} + t_{pre} + t_{mov} = t_{alarm} + t_{evac} \tag{5.31}$$

Furthermore, the ASET (t_{ASET}) is considered until untenability occurs. The untenability conditions are based on the CO (carbon monoxide) concentration and temperature [14] (see Chap. 1 for detailed information).

5.3.2 Required Data to Manage the Evacuation Procedure

The ASET should be known in order to achieve all of the potential of evacuation models for supporting emergency decision-making. It should be noted that only the comparison of required and available times can provide accurate evacuation decisions. In fact, proper decision-making should increase the margin of safety as follows:

$$t_{m\,arg} = t_{ASET} + t_{RSET} \tag{5.32}$$

Thus, we are confronted with the necessity of calculating (or estimating) the ASET. In the case of a fire, there are several models for predicting the fire dynamics, such as the fire dynamics simulator (FDS) [15]. Furthermore, these models analyse the fire characteristics, and conditions such as temperature and smoke movement, in order to calculate the time available for egress before untenable conditions are reached (ASET). These fire models can be classified into (1) field models, and (2) zone models. Field models are based on computational fluid dynamics. The computational cost and run times of these models are higher than for zone models; however, the results are more accurate. For this reason, this type of model is inappropriate for real-time calculation purposes. However, previous simulations could be run in order to obtain a database with the most likely CS (Changing Scenario) for different sources of spread patterns.

Zone models can run on any PC within a few seconds, if most of the inputs have been previously introduced. However, zone models are less accurate than field models. Likewise, other empirical or analytical models exist that might be used, despite their potential inaccuracies.

The key aspect of using any fire model that can be appropriated by its run time and computational cost is its accuracy. If we consider δ (as a percentage) as the estimated (or known) accuracy of the ASET for a specific model, then the "safety ASET" (for decision support purposes) can be obtained as follows:

$$t_{ASET_{safe}} = k_{safe} \cdot t_{ASET\,\mathrm{mod}} \tag{5.33}$$

where

$t_{ASET_{safe}}$ —ASET value for decision support purposes;

$t_{ASET\mathrm{mod}}$—ASET value obtained from the model;

$k_{safe} = 1 - \frac{\delta}{100}$ —Safety coefficient.

Based on (5.33), the margin of safety is calculated as follows:

$$t_{ASET_{safe}} = k_{safe} - t_{ASET\ \mathrm{mod}} \tag{5.34}$$

According to (5.34), any possible decision related to the evacuation process should tend to increase the margin of safety. It should be noted that the possibility of calculating the ASET depends on the type of emergency. Nowadays, in the case of a hurricane, it is possible to obtain the ASET owing to improvements in current forecasting models. On the other hand, it is highly difficult, or even impossible, to obtain the ASET for emergencies such as earthquakes, explosions, collapses, etc.

5.3.3 *Types of Decision Related to Evacuation Models*

There are several different decisions relating to emergency management that may be made using evacuation models:

1. The selection of evacuation routes.
2. Introduction of artificial delays to prevent the blockage of evacuation routes.
3. Control of the speed of evacuation.

5.3.3.1 Selection of Evacuation Routes

The RP does not always facilitate the decision-making for the selection of evacuation routes. It should be noted that two different escape routes are required for the same group of people in the RP in order to manage the possible evacuation routes.

For example, the selection of evacuation routes can be performed in complex scenarios, such as sports stadiums, high-rise buildings, shopping malls, and transport stations, which have various possible exits (provisional or definitive). We consider M as a set of possible escape routes for a group of i people, and t_{margi} as the margin of safety for each person. Thus, the selected route j should satisfy the following requirement:

$$t_{m\ \arg_j} \geq t_{m\ \arg_i} \Big| j \neq i; j, i \in M \tag{5.35}$$

where

$t_{m\ \arg_j}$ —The margin of safety for the selected evacuation route.

This is the only type of decision that can be made when the ASET is unknown. In this case, the following equation should be satisfied:

$$t_{RSET_j} \geq t_{RSET_i} \Big| j \neq i; j, i \in M \tag{5.36}$$

Or, even

$$t_{evac_j} \geq t_{evac_i} \Big| j \neq i; j, i \in M \tag{5.37}$$

5.3.3.2 Introduction of Delay Times

Another type of decision is to delay the evacuation movement of a group of people. This will allow the avoidance of possible congestion and bottleneck problems in overlapping routes. However, this is a critical decision that should be carefully considered. The key element of the problem is the insurance of a margin of safety t_{marg} greater than zero. Furthermore, these artificial delays should tend to increase the margin of safety. It should be noted that crowd speed depends on the density of people and the flow. Figure 5.5 provides an example.

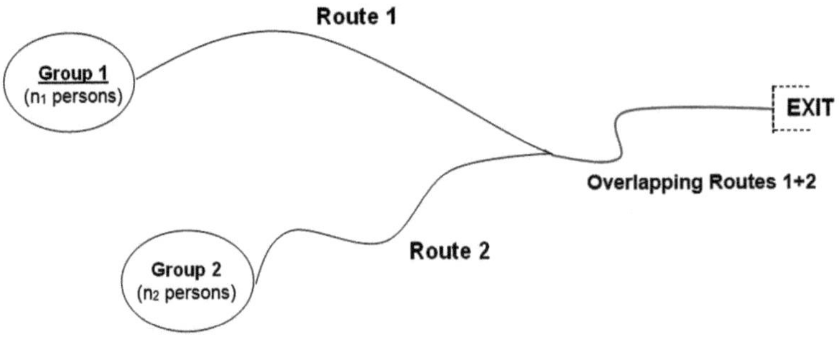

Fig. 5.5 Example of overlapping routes

We consider t_{marg1} and t_{marg2} as the margins of safety for the overlapping routes 1 and 2. Furthermore, $t_{m\arg 1}^{'}$ ($t_{m\arg 2}^{'}$) is the margin of safety for route 1 (route 2), after introducing the delay time to the corresponding group. The decision of delaying the evacuation of people is taken when the following propositional formulas are fulfilled:

$$P_1 = t_{margi} < 0 \middle| \forall i \in \{1,2\} \quad (5.38)$$

$$P_2 = t_{margi}^{'} < 0 \middle| \exists i \in \{1,2\} \quad (5.39)$$

As we can see in these propositional formulas, the decision of delaying the movement of one group of people can be made (P_0) when either one or both safety margins t_{margi} are less than zero (P_1), and both safety margins are greater than or equal to zero after delaying one group of people $t_{margi}^{'}$ (P_2).

The selection of the group of people to be delayed, and the required delay time, are complex issues that can be solved using multiple simulations. A validated evacuation model for a specific scenario should be employed to perform this type of simulation. It is convenient to employ previous simulations from a database, to ensure the requirement of real-time decision-making.

5.3.3.3 Speed Control

It should be noted that it is possible to divide an evacuation route into several parts. It is possible to control the speed of evacuation when the ASET of each piece of a route is known. Furthermore, it is possible to manage the walking speed of evacuees in each piece of a route, when continuous commands can be provided during an evacuation process. Figure 5.6 illustrates this situation.

For an egress route, the overall ASET can be calculated as follows:

$$t_{ASET} = \sum_{j=1}^{m} t_{ASET_j} \quad (5.41)$$

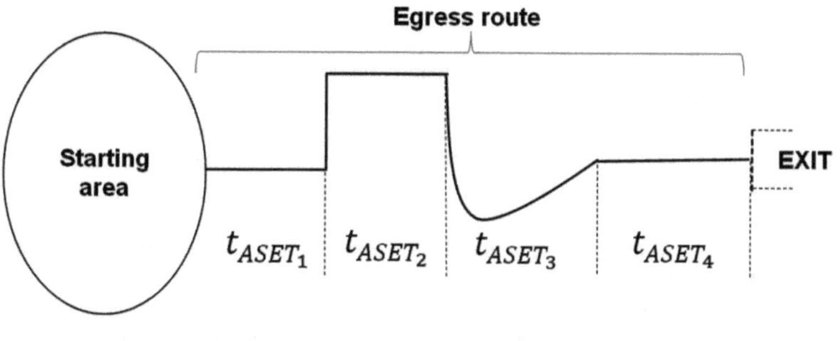

Fig. 5.6 Pieces of egress routes

where

m—The total number of parts into which the egress route is divided.

The RSET should be calculated for each piece of a route (t_{RESETj}). Thus, the following equation should be satisfied:

$$t_{m\arg_j} = t_{RSET_j} - t_{ASET_j} \geq 0 \Big| \forall j \in m \quad (5.42)$$

Based on (5.42), we can define the characteristics that should be met by the scenarios—RS and RP—in order to consider an evacuation model for decision support in emergency management. The scenarios should satisfy one of the following features:

1. The scenario should have more than one escape route.
2. At least one evacuation route should overlap for the same group of people.
3. The ASET can be obtained for sequential pieces of (at least) one escape route. Thus, the walking speeds of people can be managed with the use of commands.

5.4 Conclusions

There are many possible methods and directions for evacuation modelling in the future. Some of these may be presently unknown. This chapter has introduced some aspects that should be taken into account in a short-term vision of evacuation modelling.

Some challenges have been presented, and some preliminary solutions have been illustrated. Some tangible additional issues in evacuation modelling have not been considered in this chapter, such as changes in population, misuse of evacuation models, the verification and validation of models, and others.

It is worth saying that the development and use of evacuation modelling constitutes an element of safety science, an emergent discipline in science that must go beyond current concepts and applications.

References

1. Ronchi, E., & Kinsey, M. J. (2011). Evacuation models of the future: insights from an online survey of user's experiences and needs. In *Advanced Research Workshop Evacuation and Human Behaviour in Emergency Situations* (pp. 145–155), Santander, Spain.
2. Thompson, P. A., & Marchant, E. W. (1994). Simulex: Developing new computer modelling techniques for evaluation. In *Proceedings of the Fourth International Symposium on Fire safety Science* (pp. 613–624).
3. Thompson, P. A., & Marchant, E. W. (1995). Testing and application of the computer model SIMULEX. *Fire Safety Journal, 24*, 149–166.
4. Korhonen, T., & Hostikka, S. (2009) *Fire dynamics simulator with evacuation: FDS+Evac: Technical reference and user's guide* (VTT Working Papers 119). Espoo: VTT Technical Research Centre of Finland.

5. Fellendorf, M., & Vortisch, P. (2010). Microscopic traffic flow simulator VISSIM. In J. Barcelo (Ed.), *Fundamentals of traffic simulation* (pp. 63–94). New York: Springer.
6. *STEPS simulation of transient and pedestrian movements: User manual.* Unpublished, available with egress model from Mott MacDonald. Retrieved from http://www.mottmac.com
7. Thorton, C., O'Konski, R., & Hardeman, B. (2009). Introducting: An agent-based egress simulator. In *Proceedings of the Fourth International Symposium on Human Behaviour in Fire* (pp. 567–572).
8. Owen, M., Galea, E. R., & Lawrence, P. (1997). Advanced occupant behavioural features of the building-EXODUS evacuation model. In *Proceedings of the Fifth International Symposium on Fire Safety Science* (pp. 795–806), Melbourne, Australia.
9. Berrou, J. L., et al. (2007). Calibration and validation of the Legion simulation model using empirical data. In N. Waldau, P. Gattermann, H. Knoflacher, & M. Schreckenberg (Eds.), *Pedestrian and Evacuation Dynamics 2005* (pp. 167–181). Berlin: Springer.
10. Klüpfel, H., & Meyer-König, T. (2005). Simulation of the evacuation of a football stadium using the CA model PedGo. In S. P. Hoogendoorn, S. Luding, P. H. L. Schreckenberg, & D. E. Wolf (Eds.), *Traffic and Granular Flow'03* (pp. 423–428). Berlin: Springer.
11. Bukowski, R. W, Peacock, R. D., & Jones, W. W. (1998). Sensitivity examination of the airEXODUS aircraft evacuation simulation model. In *Proceedings of the International Aircraft Fire Cabin Research Conference* (pp. 16–20), Atlantic City, USA.
12. Gwynne, S., Galea, E. R., Lyster, C., & Glen, I. (2003). Analysing the evacuation procedures employed on a Thames passenger boat using the MaritimeEXODUS evacuation model. *Fire Technology, 39*, 225–246.
13. BSI, PD 7974-6. (2004). *The application of fire safety engineering principles to [the] fire safety design of buildings, human Factors: Life safety strategies, occupant evacuation, behaviour and condition.* London: British Standards Institute.
14. Levin, B. M. (1989). EXITT-A simulation model of occupant decisions and actions in residential fires. In *Proceedings of the Second International Symposium on Fire Safety Science* (pp. 561–570), Tokio, Japan.
15. Christensen, K., & Sasaki, Y. (2008). Agent-based emergency evacuation simulation with individuals with disabilities in the population. *Journal of Artificial Societies and Social Simulation 11*(3), 9. Retrieved from http://jasss.soc.surrey.ac.uk/11/3/9.html
16. Johnson, C. (2005). *Using computer simulations to support a risk-based approach for hospital evacuation* (Technical Report). University of Glasgow. Retrieved from http://www.dcs.gla.ac.uk/~johnson/papers/G-HES.PDF
17. Mohsen, J., et al. (2003). *Technological advances in evacuation planning and emergency management: current state of the art* (Final Report). The State University Piscataway: Center for Advanced Infrastructure & Transportation (CAIT) Civil & Environmental Engineering Rutgers.
18. Alonso, V. (2014). *Egress modelling in health care occupancies*, The Fire Protection Research Foundation, USA, Retrieved July 2014, from http://www.nfpa.org/research/fire-protection-research-foundation/
19. Kholshevnikov, V. V, &. Samoshin, D. A. (2009). Pre-school and school children building evacuation. In *Proceedings of the Fourth International Symposium on Human Behaviour in Fire* (pp. 243–254), Cambridge, England.
20. Larusdottir, A. R., & Dederichs, A. S. (2012). Evacuation of children: Movement on stairs and on horizontal plane. *Fire Technology, 48*, 43–53.
21. Ono, R., et al. (2012). Walking speed data of fire drills at an elementary school. In *Proceedings of the 5th International Symposium on Human Behaviour in Fire* (pp. 98–108), Cambridge, England.
22. Larusdottir, A. R., & Dederichs, A. S. (2012). Behavioural aspects of movement downstairs during elementary school fire drills. In *Proceedings of the 5th International Symposium on Human Behaviour in Fire* (pp. 120–127), Cambridge, England.
23. Ono, R. & Valentin, M. V. (2009). A study on evacuation of school buildings for elementary education. In *Proceedings of the Fourth International Symposium on Human Behaviour in Fire* (pp. 231–242), Cambridge, England.

24. Campanella, M. C., Larusdottir, A. R., Daamen, W., & Dederichs, A. S. (2011). Empirical data analysis and modelling of the evacuation of children from three multi-story day-care centres. *In Advanced Research Workshop Evacuation and Human Behaviour in Emergency Situations* (pp. 223–236), Santander, Spain.
25. Capote, J. A., Alvear, M. D., Abreu, O. V., Cuesta, A., & Alonso, V. (2011). Human behaviour during train evacuations: Data collected and egress modelling. In *Advanced Research Workshop Evacuation and Human Behaviour in Emergency Situations* (pp. 293–308), Santander, Spain.
26. Cuesta, A., et al. (2013). Exploring the current egress models capabilities for simulating evacuation of children through stairs. In *Proceedings of INTERFLAM 2013* (pp. 1013–1022), England.
27. Muir, H. (1991). *Human behavior in emergency situations* (Scientifics Report No. LR-HSU-15). Cranfield University for Railway Safety, England.
28. Frantzich, H. (2000). *Utrymning av tunnelbanetåg Experimentell utvärdering av möjligheten att utrymma i spårtunnel.*
29. *Research results: Passenger rail car egress time prediction.* U.S. Department of Transports, Federal Railroad Administration, RR06.
30. Norén, A. & Winér, J. (2003). *Modelling crowd evacuation from road and train tunnels—Data and design for faster evacuations* (Report No. 5127). Lund, Sweden: Department of Fire Safety Engineering, Lund University.
31. Kim, J. H., et al. (2012). Experiments of egress behavior when subway car stops on track. In *Proceedings on Human Behaviour in Fire*, Cambridge, England.
32. Fridolf, K., Nilsson, D., & Frantzich, H. (2014). The flow rate of people during train evacuation in rail tunnels: Effects of different train exit configurations. *Safety Science, 62*, 515–529.
33. Galea, E. R., & Gwynne, S. (2000). Estimating the flow rate capacity of an overturned rail carriage end exit in the presence of smoke. *Fire and Materials, 24*, 291–302.
34. Oswald, M., Schjerve, N., & Lebeda, C. (2011). Carriage evacuations in local, public rail transportation systems in case of fire experiments, findings and human behaviour. In *Advanced Research Workshop Evacuation and Human Behaviour in Emergency Situations* (pp. 309–336), Santander, Spain.
35. Peacock, R. D., Reneke, P. A., & Bukowski, R. W. (2002, December). *Fire safety of passenger trains: Phase II: Application of fire hazard analysis techniques* (NISTIR 6525). Gaithersburg, MD: NIST National Institute of Standards and Technology.
36. Kagendal, P., & Nilsson, D. (2002). Evacuation study. In *Fire safety on intercity and interregional multiple unit trains* (Report No. 5117). Lund, Sweden: Department of Fire Safety Engineering, Lund University.
37. Capote, J., et al. (2010). An evacuation model for high speed trains. In *Proceedings of the 5th International Conference on Pedestrian and Evacuation Dynamics 2010* (p. 421). Heidelberg: Springer.
38. Capote, J., Alvear, D., Abreu, A., Cuesta, A., & Alonso, V. (2012). A stochastic approach for simulating human behaviour during evacuation process in passenger trains. *Fire Technology, 48*, 911–925.
39. Markos, S. H., & Pollard, J. K. (2013). *Passenger train emergency systems: Single-level commuter rail car egress experiments*. Prepared by Volpe Center/USDOT for FRA/USDOT. Final Report. In FRA report approval process as of May 2013.
40. Galea, E., et al. (2013). The development and validation of a rail car evacuation model. In *Proceedings of Interflam 2013*, Cambridge, England.
41. Boyce, K., Shields, T. J., & Silcock, G. W. H. (1999). Toward the characterization of building occupancies for fire safety engineering: Capabilities of disabled people moving horizontally and on an incline. *Fire Technology, 35*(1), 51–67.
42. Ronchi, E., Colonna, P., Capote, J. A., Alvear, D. M., Berloco, N., & Cuesta, A. (2012). *The evaluation of different evacuation models for assessing road tunnel safety analysis, Tunnelling and Underground Space Technology*. Retrieved from http://dx.doi.org/10.1016/j.tust.2012.02.008

43. Capote, J., Alvear, D., Abreu, O., Cuesta, A., & Alonso, V. (2013). A real-time evacuation model for road tunnels. *Safety Science, 52*, 73–80.
44. Nelson, H. E., & Mowrer, F. W. (2002). Emergency movement. In P. J. DiNenno (Ed.), *SFPE Handbook of Fire Protection Engineering* (3rd ed.). Bethesda, USA: Society of Fire Protection Engineers.
45. Kuligowski, E. D., et al. (2010, November). *A review of building evacuation models* (2nd ed.) (NIST Technical Note 1680). Gaithersburg, MD: NIST National Institute of Standards and Technology.
46. Gwynne, S., et al. (1999). A review of the methodologies used in the computer simulation of evacuation from building environment. *Building and Environment, 34*, 741–749.
47. Santos, G., & Aguirre, B. E. (2005, July). *A critical review of emergency evacuation simulation models*. NIST Special Publication 1032 about the Workshop on Building Occupant Movement during Fire Emergencies.
48. Kuligowski, E. D., & Gwynne, S. M. V. (2005). What a user should know when selecting an evacuation model. *Fire Protection Engineering*, 30, 32–35, 38, 40.
49. Averill, J. D. (2010). Five grand changes in pedestrian and evacuation dynamics, In *Proceedings of the 5th International Conference on Pedestrian and Evacuation Dynamics 2010* (p. 1). Heidelberg: Springer.
50. Lord, J., et al. (2005, December). *Guide for evaluating the predictive capabilities of computer egress models* (NIST GCR 06-886). Gaithersburg, MD: NIST National Institute of Standards and Technology.
51. Gardiner, C. (2004). *Handbook of stochastic methods: For physics, chemistry and the natural sciences* (3rd ed.). Berlin, Germany: Springer. ISBN 3540208828.
52. Niederreiter, H. (1992). *Random number generation and quasi-Monte Carlo methods* (Vol. 63). Philadelphia, PA: Society for Industrial and Applied mathematics.
53. Hammersley, J. M., & Handscomb, D. C. (1964). *Monte Carlo methods* (Vol. 1). London, England: Methuen.
54. Papoulis, A. (1991). *Probability, random variables, and stochastic processes* (3rd ed.). New York: McGraw–Hill. ISBN 0-07-100870-5.
55. Grinstead, C. M., & Snell, J. L. (1997). *Introduction to probability* (2nd ed., pp. 325–360). Providence, RI: AMS Bookstore. ISBN 0821807498.
56. Gwynne, S., & Kuligowski, E. (2008). *Application modes of egress simulation, pedestrian and evacuation dynamics 2008* (pp. 397–409). Heidelberg, Germany: Springer.
57. Cuesta, A., Alvear, D., & Abreu, O. (2014). Real-time evacuation models for decision support. In *Proceedings of the 11th International Symposium of Fire Safety Science*. Christchurch, New Zealand: IAFSS (International Association of Fire Safety Science).
58. Galea, E. R., Xie H., & Lawrence, P. J. (2014). Experimental and survey studies on the effectiveness of dynamic signage systems. In *Proceedings of the 11th International Symposium of Fire Safety Science*. Christchurch, New Zealand: IAFSS (International Association of Fire Safety Science) (in press). Retrieved from http://www.getaway-project.eu/
59. Tian, J., Wang, T., Li, H., Li, L., & Want, K. (2005). DSS development and application in China. *Decision Support Systems, 4*(4), 2060–2077.
60. Papamichali, K., & French, S. (2005). Design and evaluation of an intelligent decision support system for nuclear emergencies. *Decision Support Systems, 41*(1), 84–111.
61. Alvear, D., Abreu, O., Cuesta, A., & Alonso, V. (2013). Decision support system for emergency management: Road tunnels. *Tunneling and Underground Space Technology, 34*, 13–21.
62. ISO/TR 16738: 2009. (2009). *Fire-safety engineering—Technical information on methods for evaluating behaviour and movement of people*, Geneva, Switzerland.

Glossary

Affiliation Tendency to move towards the familiar, people or places.

Alarm time The time from detection to a general alarm.

Available Safe Egress/Escape Time (ASET) Calculated time available between incident (e.g. ignition of fire) and the time at which tenability criteria area exceeded in a specific space.

Behavioural uncertainty Uncertainty associated with the current lack of knowledge in evacuation research about some of the variables affecting human behaviour of individuals and populations.

Component testing The process of checking that the components of a model work as intended.

Default setting An initial condition or algorithm provided by a developer as part of the model software.

Defend in place Life safety strategy in which occupants are encouraged to remain in their current location rather than to attempt escape during an emergency.

Detection time Time interval between incident (e.g. ignition of fire) and its detection by an automatic or manual system.

Emergent Norm Theory Individuals work together to redefine the situation and propose a new set of actions, which is the product of milling and keynoting processes.

Escape time The interval between incident (e.g. ignition of fire) and the time at which all occupants are able to reach a safe location.

Evacuation time Time interval between the time of a warning of an emergency being transmitted to the occupants and the time at which the occupants of a specified part of a building or all of the building are able to enter a place of safety. This represents the sum of pre-evacuation time and movement time.

Flow Number of people passing an area in a certain time.

Flow conditions/flow constraints The relationship between speed/flow, population density and population size.

A. Cuesta et al. (eds.), *Evacuation Modeling Trends*,
DOI 10.1007/978-3-319-20708-7

Fractional Effective Dose (FED) Ratio of the exposure dose for an asphyxiant toxicant to that exposure dose of the asphyxiant expected to produce a specified effect on an exposed subject of average susceptibility.

Habituation A process has become routine in nature through continuous exposure over a period of time.

Hypervigilance A particular sensitivity to the existence of certain cues leading to elevated perception of risk.

Incapacitation Inability to take effective action to accomplish one's own escape from a danger (e.g. a fire).

Incident A sudden, usually unexpected occurrence of a danger demanding immediate remedy or action.

Intrinsic uncertainty Uncertainty associated with the physical and mathematical assumptions and methods that are intrinsic to the model formulation.

Keynoters Individuals who advance suggested interpretations of the event or suggestions on what do to next.

Local movement Simulation of space navigation intended as the representation of the interactions of individual agents with the environment.

Measurement uncertainty Uncertainty is associated with the experimental measurement itself.

Milling A communication process whereby individuals come together in an attempt to define the situation, propose and adopt new appropriate norms for behaviour, and seek coordinated action to find a solution to the shared problem at hand.

Model input uncertainty Uncertainty associated with the parameters obtained from experimental measurements that are used as model input.

Movement Time/travel time The time required for evacuees to walk to a safe place. A safe place can be an exit or an escape route. This time also includes the time required to flow through exits and escape routes.

Normalcy bias The assumption that nothing unusual is taking place even when faced with unusual external cues.

Optimism bias The assumption that no personal threat exists.

Pathfinding The simulation of route choice in time or space.

People density The number of persons divided by the available escape route area pertinent to the space where the persons are located.

Pre-evacuation time/delay time/pre-movement time/pre-travel activity time/waiting time The interval between the time at which a warning of an emergency is given and the time at which the first move is made by an occupant towards an exit. This consists of two components: recognition time and response time (defined below).

Note:

For groups of occupants, two phases can be recognized:

- The pre-travel activity time of the first occupants to move.
- The pre-travel activity time distribution between the first and last occupants to move.

Protective Action Decision Model (PADM) A framework that describes the information flow and decision-making that influences protective actions taken in response to natural and technological disasters.

Protective actions Actions that people take to protect themselves, others and/or property in an emergency.

Recognition time The interval between the time at which a warning of an emergency is given and the first response to the warning.

Required Safe Egress/Escape Time (RSET) Time required for escape. Calculated time period required for an individual occupant to travel from their location at the time of incident (e.g. ignition of fire) to a safe refuge.
Note:
RSET represents the sum of detection time, warning time, pre-travel activity time and travel time.

Response time The interval between the time at which the first response to the event occurs and the time at which travel begins to a safe location.

Risk perception The personalization of the risk related to a current emergency event.

Route availability The routes available to the evacuees.

Route usage/route choice/path selection The routes selected by the evacuees from those available.

Satisficing Is a method in which an individual chooses the first option that seems to work, though not necessarily the best option overall.

Social influence Changes in attitudes, beliefs, opinions or behaviour as a result of the fact that one has encountered others.

Steering behaviour The physics of the agent local movement based on behaviourally determined steering forces which are applied to the mass of each agent at each time-step.

Travel speed/walking speed/movement speed The maximum uncongested walking speed at which individual evacuees move towards a place of safety.

MIX
Papier aus verantwortungsvollen Quellen
Paper from responsible sources
FSC® C105338

If you have any concerns about our products,
you can contact us on
ProductSafety@springernature.com

In case Publisher is established outside the EU,
the EU authorized representative is:
Springer Nature Customer Service Center GmbH
Europaplatz 3, 69115 Heidelberg, Germany

Printed by Libri Plureos GmbH
in Hamburg, Germany